Explanations

Explanations

Styles of Explanation in Science

Edited by

John Cornwell

JESUS COLLEGE, CAMBRIDGE

OXFORD
UNIVERSITY PRESS

OXFORD
UNIVERSITY PRESS

Great Clarendon Street, Oxford OX2 6DP

Oxford University Press is a department of the University of Oxford.
It furthers the University's objective of excellence in research, scholarship,
and education by publishing worldwide in

Oxford New York

Auckland Bangkok Buenos Aires Cape Town Chennai Dar es Salaam
Delhi Hong Kong Istanbul Karachi Kolkata Kuala Lumpur Madrid Melbourne
Mexico City Mumbai Nairobi São Paulo Shanghai Taipei
Tokyo Toronto

Oxford is a registered trade mark of Oxford University Press
in the UK and in certain other countries

British Library Cataloguing in Publication Data
Data available

ISBN 0 19 860778 4

1

Typeset by Footnote Graphics Limited, Warminster, Wiltshire
Printed in Great Britain by Clays Ltd., St Ives plc

'I wish he would explain his explanation'

LORD BYRON
OF SAMUEL TAYLOR COLERIDGE
IN *DON JUAN*

Preface

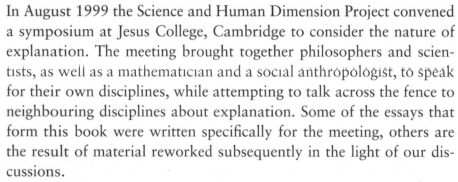

In August 1999 the Science and Human Dimension Project convened a symposium at Jesus College, Cambridge to consider the nature of explanation. The meeting brought together philosophers and scientists, as well as a mathematician and a social anthropologist, to speak for their own disciplines, while attempting to talk across the fence to neighbouring disciplines about explanation. Some of the essays that form this book were written specifically for the meeting, others are the result of material reworked subsequently in the light of our discussions.

The symposium was made possible by the generosity of the Mrs L. D. Rope Third Charitable Settlement and the encouragement of Mr Crispin Rope. Finally, I must thank the Master and Fellows of Jesus College, Cambridge for their enthusiastic support.

J. C.
Cambridge
March 2003

Contents

Contributors

Professor Peter Atkins Lincoln College, Oxford OX1 3DR, UK.
peter.atkins@lincoln.ox.ac.uk

Professor John D. Barrow Department of Applied Mathematics
and Theoretical Physics, Centre for Mathematical Sciences,
Cambridge CB3 0WA, UK.
j.d.barrow@damtp.cam.ac.uk

Professor Tian Yu Cao Philosophy Department, Boston University, USA.
tycao@bu.edu

John Cornwell Jesus College, Cambridge CB5 8BL, UK.
jc224@cam.ac.uk

Professor Jack Goody St John's College, Cambridge CB2 1TP, UK.
jrg1@hermes.cam.ac.uk

Dr David Hanke Department of Plant Sciences, Downing Street,
Cambridge CB2 3EA, UK.
deh1000@cam.ac.uk

Professor Peter Lipton Department of History and Philosophy of Science, Free School Lane, Cambridge CB2 3RH, UK.
peter.lipton@kings.cam.ac.uk

Professor Colin McGinn Rutgers, the State University of New Jersey, 26 Nichol Avenue, New Brunswick, NJ 08901-1411, USA.
cmcginn@email.msn.com

Professor Martin Rees Institute of Astronomy, Madingley Road, Cambridge CB3 0HA, UK.
martin.rees@kings.cam.ac.uk

Professor Steven Rose Department of Biological Sciences, The Open University, Milton Keynes MK7 6AA, UK.
s.p.r.rose@open.ac.uk

Professor William C. Saslaw Department of Astronomy, PO Box 3818, Charlottesville, VA 22903, USA.
wcs@virginia.edu

Dr Jon Turney Penguin Press, 80 Strand, London WC2R 0RL, UK.
jonturney@penguin.co.uk

Professor Steven Weinberg Physics Department, University of Texas, Austin, TX 78712, USA.
weinberg@physics.utexas.edu

John Cornwell

Introduction

NOWHERE, perhaps, are different sorts of explanations more crucial and significant in their consequences for everyday life than in courts of law, where witnesses, defendants, lawyers, judges, and juries attempt to disentangle versions of the how and the why with attendant evidence and argument. Increasingly, moreover, litigation is the context in which scientific explanations make connections with our working and domestic lives, our liabilities, our attempts to understand human nature, and how and why we do the things we do. Hence, in a book of essays that seeks to explore the nature and validity of varying styles of explanation, from the vantage point of different academic disciplines mainly within the natural sciences, it is appropriate to introduce and illustrate the scope of our theme with a memorable civil action in which reductionist scientific explanations were in competition with others, including social, psychological, and even metaphysical ones.

In September of 1989, Joseph T. Wesbecker, a 47-year-old printer, returned to Standard Gravure, his former place of work in Louisville, Kentucky, and shot 20 of his co-workers, killing 8 and injuring 12, before committing suicide in front of the plant supervisor's office. It was soon discovered that Wesbecker had been on a course of the anti-depressant, Prozac, for depression. So it was that Eli Lilly of

Indianapolis, the pharmaceutical company responsible for making and distributing the drug, became the focus of a civil liability suit brought by the survivors, many of whom were permanently disabled, and the relatives of the dead. The case went to a jury trial in Louisville, Kentucky in September 1994, and lasted three months.

The court proceedings were unusual because they invoked a variety of explanations for Wesbecker's behaviour, not least the action of chemicals, synthetic and natural, on the human brain, in sickness and in health; profound issues of personal responsibility in the light of new brain science, and traditional psychiatry; acceptable procedures for clinical trials of drugs, and the ethics associated with drug package warnings. Taking place at the mid-point of 'The Decade of the Brain', an epoch nominated by a joint resolution of the House and the Senate of the United States as special to neuroscience, there were many expert witnesses for both the defence and the plaintiffs. The air was thick with competing causal, descriptive, and metaphysical explanations, invoking medical, scientific, social, commercial, and even, outrageously and absurdly, 'quantum physics'.

At one stage, during the cross-examination of Dr Leigh Thompson, head of research at Eli Lilly, the question of causal explanations arose. The plaintiff's lawyer Mr Paul Smith asked whether 'Prozac can cause violent aggressive behavior'. To which Dr Thompson replied: 'I would say that almost anything is possible, including the possibility that this courthouse will move across the street to the park outside. In quantum mechanics, that's clearly possible.'

Thompson's introduction of quantum mechanics into the argument caused a minor sensation in the court, even startling the judge, who normally remained stony-faced. Calling the lawyers to the bench, the judge wanted to discuss this intervention of physics a little more closely. 'I couldn't even *say* quantum physics or whatever it was, quantum mechanics', said one of the lawyers.

'Actually,' said the judge, 'I'm a kind of physics buff or whatever, and he's technically absolutely correct.'

'It is correct', said Lilly's lawyer, nodding sagely. 'Yeah.'

The judge, however, had not seemed to grasp that the witness had introduced quantum physics as an outrageous piece of obfuscation, a ploy, and had got away with it.[1]

The lawyers on both sides, quite naturally, were seeking accounts likely to aid their arguments: their explanations from the very outset were unlikely to be compatible.

The purpose of jury verdicts in civil actions, however, is to involve members of the community in decisions about how people should behave towards each other. In consequence there were wider questions, remoter explanations, involving competitive business practices in contracting markets, endemic workplace stress, the American mania for civil-liability suits, and the allure of high-stakes litigation. There were issues, moreover, involving the gulf between authentic public-health needs and the commercial goals of the pharmaceutical industry; the public's right to know the unadorned truth about medication, and the pharmaceutical industry's tendency to withhold selective information in the interest of corporate aims.

According to his psychiatrist, Wesbecker had been prescribed Prozac to alleviate depression related to workplace stress and his complaints of alleged unfair treatment by the management at Standard Gravure. Plaintiffs' counsel argued that the drug had disrupted Wesbecker's 'impulse control' to a point where he was not responsible for his actions. Lilly, he insisted, had knowingly manufactured and marketed a drug that caused agitation in certain individuals, and had failed to provide the appropriate package warnings in the United States.

Lilly denied negligence in their testing and marketing of Prozac. Having explained the reductionist neuropsychological description of depression the company routinely employed to promote Prozac (lowered activity of serotonin in the brain), the defence strategy focused on Wesbecker as a product of a dysfunctional family that had suffered from hereditary mental illness in three generations.

But when the defence's case began to founder under the weight of contradictory explanations, their lawyers constructed an argument based on Wesbecker's status as a free individual. The killings, they said, were wholly caused by Wesbecker's unimpaired free will. Here was the ultimate explanation for the whole affair: he was a bad man, and he did what he did because he wanted to. The circumstance of this final explanation aptly describes a familiar and perhaps incompatible tension in American society today – between forms of medicalized neurogenetic determinism, on the one hand, and notions of individual moral freedom espoused, for example, by religionists of every persuasion, on the other.

Since Wesbecker was a rational individual capable of making plans and choices, argued Lilly's trial lawyer, Ed Stopher, his ability to control his actions had been affected neither by chemical intrusion nor by the influence of nature and nurture. Wesbecker had a personality disorder, defence expert witness Dr Granacher declared, but he had 'intent', he had 'goals', and he was 'in control'.

It would be easy to dismiss the final apparently incompatible accounts – chemical determinism versus untrammelled free will – as mere tactics in a civil dispute. After all, the goal of the plaintiffs was not so much to establish the exhaustive 'truth' of a complex set of events, as to assign blame and put a dollar value on the resulting damages. It is surely significant, however, that these versions of who, or what, was responsible for Wesbecker's actions were wholly lacking in a social dimension. The litigants' denial of Wesbecker's existence as an individual shaped by his relationships with his family, neighbours, and co-workers bears the unmistakable hallmark of reductionist thinking, the scientific methodology by which entities are described in terms of their smallest material parts.

Reductionist explanations – explaining nature by reducing phenomena right down to the workings of atoms and molecules – have been decisively successful for Western science and technology since the second half of the nineteenth century. As an account of human identity and society they have remarkable shortcomings. The tendency to attempt explanations of complicated things and processes by simple and basic principles starts as early as 600 BC with Thales of Miletus, who taught that everything is made of water, an appealing notion that was unlikely to go anywhere since it was not true. Two thousand years on, in the seventeenth century, however, Galileo and Newton described an atomized, mechanistic world in which causal relationships could be understood and predicted in terms of the physics of falling objects, the motion of projectiles, tides, and the heavenly bodies. The idea that biological organisms, too, were part of a mechanistic world picture was developed by René Descartes, the seventeenth-century philosopher and mathematician, who believed that animals were machine-like automata working on hydraulic principles. Descartes eventually extended his mechanistic model to human nature itself, thus creating one of the most troubling legacies of the Enlightenment. In our bodily functions, Descartes maintained, we are mechanistic and in all respects similar to the animal kingdom, and yet we have consciousness, rationality, and free will – faculties that reside not in the material world but in the realm of the spirit. He therefore posited the existence of two sorts of stuff, *res extensa* and *res cogitans*: material stuff, which can be observed and measured, and 'thinking' stuff, the stuff of the mind, which lies beyond scientific investigation, and hence outside of nature. The frontier of body and soul, according to Descartes, was located in the brain, at the pineal gland. Thus the new sciences of physics and biology were from the outset disconnected from a theory of human freedom and responsibility.

The distinction between body and soul, widely known as 'dualism', has gone through many variations since the seventeenth century. And yet dualism, of the kind expounded by Descartes, remains the inevitable consequence of a belief that would say, with Francis Crick, 'You're nothing but a pack of neurones.' For in order to have a human identity composed of both chemical determinacy and moral agency, it is necessary to conjure up, as did Descartes, a will that is as removed from the dynamics of a social context as it is removed from the assemblies of molecules and neurones.

The survival of Cartesian body–soul dualism into the twenty-first century as an explanation of human behaviour is a token of its extraordinary resilience, not least its potential to reduce people to exploitable mechanical objects, while nevertheless insisting that they remain wholly responsible for their actions. Hence dualism continues to create a rationale for treating people as machines while insisting that they possess an incorporeal individual freedom that transcends not only deep-seated personality disorders but familial, workplace, and societal pressures. The social failure of such a model of human nature, with its attendant explanations for behaviour, is its encouragement of a social climate advantageous to self-seeking individuals with a superficial humanism.

Ironically, for Lilly, whose fortunes were increasingly built on the premise of explanations for human behaviour based on brain chemistry, it was precisely the psychopharmacological notion that there could be a single chemical basis for aggression (lowered activity of serotonin, again) that had encouraged the plaintiffs to go to court in the first place. In other words, the application of reductionist explanations to human nature had emboldened the sort of litigation that habitually denies the dynamics of social responsibility in liability suits.

As it turned out, the defence counsel started by constructing a case built on nature–nurture determinacy, drawing on four hundred depositions of Wesbecker's sorry life and background. These documents should have provided ample material for a narrative of Wesbecker's social existence, portraying him in the complex context of the workplace, family, neighbourhood, with a cluster of explanatory approaches. But Lilly's lawyers chose to concentrate on a narrowly selective version of Wesbecker's hereditary and familial background in order to show that his murderous actions were wholly 'inevitable'. At the same time, in their background defence of Prozac, they reinforced a rigorously reductionist view of human identity summed up by the reflection of an expert witness, the late Dr Ray Fuller (joint discoverer of fluoxetine hydrochloride, or Prozac) that 'behind every crooked thought there lies a crooked molecule'.

The latter line of argument was hardly surprising since the future of psychopharmacology, as revealed in the claims made for Prozac, rests on the premise that individual deviancy and unhappiness – leading to social disorder – are nothing more than specific, identifiable, and controllable chemical software programs running in the brain. Hence the role of the pharmaceutical manufacturer is to discover the links between biochemistry and behaviour, and then produce appropriate pharmacological 're-write' antidotes. Significantly, however, Lilly was reluctant to offer any explanation, still less apology, for the fact that Prozac had failed to help the sort of patient it had been designed for. And having failed to construct a case based on genetic and environmental determinism, they appealed to 'free will', which arrived, like the US cavalry, in the nick of time.

*   *   *

Our lives, state of health, relationships, behaviour, and experience of the natural world, and the technologies that shape our contemporary existence, are subject to a superfluity of competing, multifaceted, sometimes incompatible explanations. Explanation is the stuff of family life and quarrels, the diet of soaps that entertain us on television ('Why on earth did you do that?'), the vicissitudes of economics and the politics, the fundamentalisms of religion, the philosophy and logic that underpins ethics, and the forces that drive our quests to understand nature and the universe: from quantum physics to the Big Bang. The Science and Human Dimension Project at Jesus College, Cambridge, founded in 1990 to contribute to public understanding of science, reached a stage, at the end of its first decade, when it seemed a matter of urgency to explore the mechanisms, validity, and connections and disparities, between various forms and styles of explanation that we employ in scientific and philosophical enquiry, and which land, invariably enfeebled and distorted, in our exchanges of everyday life, mostly via the media. The Science and Human Dimension Project, with its special interest in how the science media – book publishing, TV documentary, periodicals – report science, had repeatedly encountered misapprehensions, ambiguity, and general befuddlement, in characterizing and reporting different styles of explanation; the more so, when a pluralism of explanations was invoked to enquire more deeply into a theme, a theory, a phenomenon, or a description of what scientists are doing. In no case had this been more obvious than in the project's meeting on the vexed topic of consciousness, held at Cambridge in August 1998,[2] when a symposium of philosophers, neuroscientists, physicists, psychologists, a novelist, and two theologians, went at the topic from their different perspectives, promoting, by inference, an extraordinarily variegated cluster of explanations.

In August 2000, therefore, we convened a meeting of philosophers and scientists, as well as a mathematician and a social anthropologist, to speak for their own disciplines, while attempting to talk across the fence to neighbouring disciplines about explanation. The result is this book, whose contributions continued to mature beyond the meeting itself until going to press in the spring of 2004.

We were extremely fortunate to have as a kind of guardian angel in this enterprise the assistance of the philosopher Professor Peter Lipton, of the Cambridge University Department of History and Philosophy of Science, who is the author of *Inference to the best explanation*. Professor Lipton's contribution is cast at the profound level of seeking answers about the 'goods of explanation', by which he means the quest for some kind of explanation of explanation itself. Such a quest, he insists, involves 'an independent account of understanding'. So he begins by considering different accounts of what understanding amounts to, settling, in the final analysis, on a 'causal view'. But why, he then asks, should we consider causes more explanatory than effects? The answer, he suggests, is that 'only causes can make the relevant difference between the occurrence and non-occurrence of the thing we want explained. But there is another 'explanatory good', he goes on: 'not a type of understanding, but what understanding, especially causal understanding, is good for'. It is good, he tells us, 'for causal inference'. In his conclusion, Professor Lipton expands on the function of explanation as inference. A function, he points out, may well be an effect, but he argues that function explanations are nevertheless causal, not 'effectal'. In biology, for example, explaining the presence of a trait functionally can be to inform about the evolutionary aetiology of a trait. To say that the function of a polar bear's white fur, he writes, 'is camouflage is to explain the presence of fur of that colour in terms of a causal history of evolution in which the possession of such fur by earlier bears on their progenitors conferred a selective advantage and so caused there to be later bears with the same traits.'

In the realm of explanations that derive from the natural sciences, physics tends to pull rank on its rival disciplines. The eminent physicist Steven Weinberg has also distinguished himself in popular accounts of theories of everything, for example his Dreams of a Final Theory, and how his discipline situates itself in relation to philosophy of science. His contribution, 'Can science explain everything? Anything?' is at once historical and analytical, addressing the limits of explanation in physics and at the same time clarifying what physics does not, and cannot, attempt to explain.

Sir Martin Rees, cosmologist and Astronomer Royal, connects with Steven Weinberg by addressing the limited sense in which some sciences can claim a status that, although not superior, is distinctive. 'As Steven Weinberg has emphasized, if you go on asking why? why? why? you end up with a fundamental question either in particle physics or cosmology.' Rees argues that we seek a 'final theory' not because the rest of science (or even the rest of physics) depends on it but because it engages with deep aspects of reality. In this sense cosmology, he claims, is a 'special' science.

Turning towards more metaphysical explanation, Rees addresses the absorbing topic of the anthropic principle, or the apparent 'fine tuning' of the universe. He rejects in short order coincidence, and just as swiftly 'Providence', the idea of a creator on the basis of a designer God, as espoused by thinkers from William Paly in the eighteenth century to John Polkinghorne. Rees then runs a version of the 'multiverse' theory in which an infinity of separate universes would explain the emergence of one in which the conditions were just right for a planet that brought forth life. From this point he expands into an 'epistemological' discussion of the term universe that leads to what he calls 'a key conceptual innovation' – the realization that our universe may be 'vastly larger than the domain we can now (or, indeed, can ever) observe. Anthropic explanations, he concludes, will then be the best we can hope for, and cosmology will in some respects resemble the science of evolutionary biology.

Steven Weinberg remarks in his contribution that philosophers are trying to develop an answer to the question 'What is it that scientists do when they explain something?' by looking at what scientists are actually doing when they *say* that they are explaining something. Professor William Saslaw, an astrophysicist who works at the University of Cambridge, and at the University of Virginia, clarifies for us some of the basic functions of working physicists, starting with the assertion that physicists have refined the technique of asking questions they know they can answer. 'The artistry and imagination of physics', he maintains, 'consists in finding these questions and the appropriate methods for their solution.' From these beginnings, Saslaw leads us through the stages that culminate in what a physicist would describe as 'an elegant explanation'. First he discusses this notion of elegance in the abstract; then he provides two examples – one in classical physics, the other in the recent study of cellular automata – of self-organizing systems.

Saslaw starts by citing Bertrand Russell, who said that physics differs from mathematics, in which 'we never know what we are talking about, nor whether what we are saying is true.' John Barrow, who holds the Chair of Public Understanding of Mathematics at Cambridge, sets out to reveal the ubiquity of mathematics as a means of explaining the world around us, its 'unbounded character' and its 'disconcerting incompleteness'. Barrow goes on to describe the utility of mathematics, as a result of quantization, upwards from elementary particles, and the mystery of mathematical explanation in its 'uncanny foresight'. So is mathematics an 'entirely human' creation? Barrow comments that, if this were true, it is strange that mathematics 'becomes better and better as we go farther and farther away from the arena of local human experience and the realm in which the process of natural selection fashioned our mental categories.'

Looking more closely at this suggestion of the Platonic nature of mathematics, Barrow considers the close relationship that once existed between theology and mathematics in their parallel quests for absolute truth. How could a theologian know anything about the nature of God? 'Their justification was in the success of Euclid's geometry. It was the prime example of our success in understanding a part of the ultimate truth of things. And if we could succeed there, why not elsewhere as well?' In time, however, the philosophical status of Euclidean geometry was undermined, leading to a variety of 'forms of relativism about our understanding of the world', the notion that mathematics was much bigger than physical reality, and a proliferation of mathematical modelling. Mathematical experience parted company with a physical existence.

But was mathematics complete? At the World Congress of Mathematics in 1900 the distinguished German mathematician David Hilbert proposed a great challenge for the twentieth century: to prove that mathematics was complete and consistent. It took thirty years to reveal, within Kurt Gödel's famous proof, that no non-trivial axiomatic system can be both complete and consistent – a result that astonished the world and continues to reverberate to this day. One of the interesting things about this deduction, according to Barrow, is to ask whether it places real limits on scientific understanding of the universe. 'Gödel's monumental demonstration, that systems of mathematics have limits, gradually infiltrated the way in which philosophers and scientists viewed the world and our quest to understand it.' Barrow goes on to cite the theologian–philosopher Stanley Jaki, as believing 'that Gödel prevents us from gaining an understanding of the cosmos as a necessary truth.' Which, of course, carries interesting theological and philosophical implications. By the same token, however, Gödel's insights have given rise to a different way of viewing the implications for physics – you can never know whether your ultimate theory, or explanation, is the ultimate theory or not.

In contrast to these journeyings towards the ultimate, Peter Atkins, the Oxford chemist, invites us to explore explanation in the ambit of 'ponderable matter'. He starts with the work of Antoine Lavoisier in the eighteenth century: 'when he brought the chemical balance to bear on matter, he turned a qualitative body of knowledge into a physical science.' In this crucial step, Atkins points out, alchemy became chemistry, numbers could be attached to matter, and the rigour of quantitative scientific investigation used to tease out understanding. From this point, and over the next two centuries, it was the chemists who explained, or 'gathered insight' into, the nature of matter. 'Dalton justified atoms, Rutherford discerned something of their structure, thermodynamics, rationalized relationship between the bulk properties, and quantum mechanics rendered the understanding of atoms precise.'

Atkins takes us through a remarkable 'brief history' of chemistry which is nothing less than a 'brief history' of reductionist explanation in science. Thence Atkins works towards his conclusion that, although life has to be understood in the light of the Second Law as the great unwinding of the cosmos, 'we can understand its richness by recognizing that there are profound kinetic constraints on the rates at which reactions take place ... Life is life because of the complexity of the network of reactions that mediate and retard the Great Unwinding.'

Complexity is the theme of Professor Stephen Rose's contribution, which insists on the importance of cultural as well as natural perspectives in explanations of human behaviour. Rose has been a doughty combatant in his books and generally in the media against simplistic reductionism as applied to human behaviour, and he uses the opportunity of his essay to draw up a list of important principles for 'the biology of the future, and the future of biology'. Scientific knowledge, he declares, is not absolute but provisional, being socially, culturally, and technologically and historically constrained; in consequence, science at the level of qualitative conclusions about nature, and about our own nature in particular, is essentially pluralist. Explanations in different disciplines, he emphasizes, deal with different levels of organization of matter and we run into trouble as soon as we apply qualities or concepts in one domain that are more suited to another. 'Thus genes', he reminds us, 'cannot be selfish; it is people, not neurones nor yet brains or minds, who think, remember, and show emotion.' Causes, he goes on, are multiple, and phenomena richly interconnected. Living organisms exist in four dimensions; organisms and environment interpenetrate; and organisms are active players in their own destiny.

Dr David Hanke, meanwhile, pleads for the avoidance of explanations in development that are 'coloured by the prejudices of the prevailing culture', in particular the assigning of teleological or purposeful intentions to animal and plant behaviour. As an example he takes commentaries on development in plants in relation to light and darkness, noting that we live in a culture which regards light as positive and darkness as negative. He criticizes, in particular, the tendency of Richard Dawkins, among others, to justify the use of teleological reasoning as virtually harmless and as an aid to thinking. But Hanke thinks this is 'lazy and wrong', because it is a 'straitjacket for the mind'. He believes that it is time to 'stop expecting that any part of the living world can be defined by function, however seductive or merely comfortable that feels.'

The subsequent two contributions, by Professors Colin McGinn and Tian Yu Cao, offer the perspective of philosophy on explanation from two different standpoints. McGinn is a philosopher of mind, and Yu Cao has been working in field theory and physics. McGinn, who is interested in the mind–body problem and consciousness, explores the relationship between objective and subjective realms, testing the validity of classic type-identity materialism. In a closely argued technical essay he concludes that the correct view of the mind–body relation is left open. 'All I have contended is that the usual kinds of materialistic identity theory are committed to an unacceptably subjective conception of the physical world.' In this sense McGinn is suggesting important limits to explanation within discussion of the problem of consciousness.

Tian Yu Cao, on the other hand, takes an 'ontological' approach to explanation in the domains of quantum physics and space-time, as well as consciousness, the origins of life, and the origins of the universe. In a commentary reflecting on the reductionism of Weinberg, and the dualism of Putnam, Tian Yu Cao proposes that his ontological approach to explanation offers a powerful reconciliation of unity and diversity of nature. 'In terms of entity,' he writes, 'the ontological approach agrees with reductionism that there is no unbridgeable gap between entities at different levels that cannot be closed by causal connections. Entities at a higher level are always causally explainable by the behaviour entities at an underlying level. Thus the unity of nature can be argued on the basis of causal connectedness of entities at adjacent levels, and thus at all levels.'

The final two chapters appropriately compare and contrast 'scientific' explanations with the other kinds of explanation. Jack Goody, the anthropologist, provides a useful survey of the history of the two broad approaches in the social sciences – scientific, or deterministic, explanations, and interpretations that take into account the phenomenology of free, reflexive human subjects. He concludes that within his own discipline 'we see a general tendency to move from explanation to interpretation partly on the grounds that the human sciences are different. Certainly we need to try to understand the actor's point of view, the meaning of social action to him or herself.'

Jon Turney, who is a media sociologist specializing in science and education, explores explanation as narrative in the qualitative conclusions of science as a literary genre and as a form of teaching. He takes us through the work of John Ogborn and his colleagues at the Institute of Education in London, showing how in classroom teaching explanation is seen as a special kind of story-telling. These narratives include protagonists with their own capabilities; members of the cast enact one of the many series of which they are capable; and these events have a consequence which follows from the nature of the protagonists and the events they enact. Parallel with this approach Turney cites Brian Green's *Elegant universe*, in which Green employs, for example, the metaphor of violin – vibrating, oscillating, and resonating – to explain string theory. These approaches, concedes Turney, may be 'shadowy', but he concludes that it is 'difficult to see where else a writer eschewing mathematics could turn for a way of approaching these strange objects.'

Notes

1. J. Cornwell. *Power to harm* (London, 1996), pp. 158–9.
2. J. Cornwell (ed.). *Consciousness and human identity* (Oxford, 1998).

Peter Lipton

What good is an explanation?

Introduction

WE are addicted to explanation, constantly asking and answering 'why' questions. But what does an explanation give us? I will consider some of the possible goods, intrinsic and instrumental, that explanations provide.

The name for the intrinsic good of explanation is 'understanding', but what is this? In the first part of this paper I will canvass various conceptions of understanding, according to which explanations provide reasons for belief, make familiar, unify, show to be necessary, or give causes. Three general features of explanation will serve as tests of these varied conceptions. These features are:

(1) the distinction between knowing that a phenomena occurs and understanding why it does;

(2) the possibility of giving explanations that are not themselves explained;

(3) the possibility of explaining a phenomenon in cases where the phenomenon itself provides an essential part of the reason for believing that the explanation is correct.

There are many other aspects of our explanatory practices that a good account of explanation and understanding should capture, but these simple tests provide surprisingly effective diagnostic tools for the evaluation of broad conceptions of the nature of understanding. It will turn out that the causal conception of understanding does particularly well on the tests, though of course it too faces various difficulties. The balance of this essay focuses on the causal conception. After addressing some of the difficulties it faces, I will ask why causes explain. Why, in particular, do causes rather than effects explain? One possible answer is that causes 'make the difference' between the occurrence and non-occurrence of what they explain. Several features of our explanatory practices will be adduced to evaluate this hypothesis. In the final section, I will consider an instrumental good of explanation revealed by the account of Inference to the Best Explanation. Explanation is an important route to the discovery of causes. This allows a functional explanation of explanation, according to which the question 'Why do causes explain?' may itself have a causal answer.

Three features of explanation

There are some simple and relatively uncontroversial features of explanation that can be used to test conception of understanding. I will use the three I have just listed. The first of these is the **gap between knowledge and understanding.** Knowing that something is the case is necessary but not sufficient for understanding why it is the case. We all know that the sky is sometimes blue, but few of us understand why. Typically, when people ask questions of the form 'Why P?', they already know that P, so understanding why must require something more than knowing that. If one's aim is to get a grip on the goods that explanations provide, it is useful to ask: what more than knowledge does understanding require? And if an account of understanding is unable to make the distinction between knowing that and understanding why, it is a bad account.

The second test feature is the **why regress**. As most of us discovered in our youth, and to our parents' consternation, whatever answer someone gives to a why question, it is almost always possible sensibly to ask why the explanation itself is so. Thus there is a potential regress of explanations. If you ask me why the same side of the moon always faces the earth, I may reply that this is because the period of the moon's orbit around the earth is the same as the period of the moon's spin about its own axis. This may be a good explanation, but it does not preclude you from going on to ask the different but excellent question of why these periods should be the same. For our purposes, the salient feature of the why regress is that it is benign: the answer to one why question may be explanatory and provide understanding even if we have no answer to why questions further up the ladder. This shows that understanding is not like some substance that gets transmitted from explanation to what is explained, since the explanation can bring us to understand why what is explained is so even though we do not understand why the explanation itself is so. Any account of understanding that would require that we can only use explanations that have themselves been explained fails the test of the why regress.

The final feature that I will use to test conceptions of understanding is the phenomenon of what are known as **self-evidencing explanations** (cf. Hempel 1965, 370–4). These are explanations where what is explained provides an essential part of our reason for believing that the explanation itself is correct. Self-evidencing explanations are common, in part because we often infer that a hypothesis is correct precisely because it would, if correct, provide a good explanation of the evidence. Seeing the disembowelled teddy bear on the floor, with its stuffing strewn throughout our living room, I infer that Rex has misbehaved again. Rex's actions provide an excellent if discouraging explanation of the scene before me, and this is so even though that scene is my only direct evidence that the misbehaviour took place. To take a more scientific and less destructive example, the velocity of recession of a galaxy explains the red shift of its characteristic spectrum, even if the observation of that shift is an essential part of the scientist's evidence that the galaxy is indeed receding at the specified velocity. Self-evidencing explanations exhibit a kind of circularity: H explains E while E justifies H. As with the why regress, however, what is salient is that there is nothing vicious here: self-evidencing explanations may be illuminating and well supported. Any account of understanding that rules them out is incorrect.

Five conceptions of understanding: reason, familiarity, unification, necessity, and causation

We now have three important features of explanation: there is a distinction between knowing that and understanding why, the why regress is benign, and good explanations may be self-evidencing. Armed with these test features, I want now to consider five broad conceptions of understanding, conceptions of what intrinsic goods an explanation provides. The first two of these conceptions – reason and familiarity – make understanding fundamentally an epistemic matter; the last two – necessity and causation – make it metaphysical or ontological. The middle conception – unification – can go either way, depending on how it is itself analysed. These conceptions of understanding are not mutually exclusive, because different explanations could provide different types of understanding and because a single explanation could yield more than one good.

First, we have the **reason** conception of understanding. Understanding is here identified with having a good reason to believe. We understand why something occurred when we have good reason to believe that it did in fact occur, and this good reason is just what an explanation provides (cf. Hempel 1965, 337, 364–76). This view has some attractions. When we ask why questions, sometimes what we really want is not an explanation but a reason for belief. Here 'Why P?' is short for 'Why should I believe that P?' The reason conception provides a unitary account of explanation seeking and reason seeking why questions: both are actually reason seeking. Indeed the word 'reason' itself has just this ambiguity: it may mean either reason for belief or reason why. Another attraction of the reason conception of understanding is negative: it avoids any dubious metaphysical notions, and relies on a notion – reason for belief – that we must appeal to in any case if we are to do epistemology at all.

Now for the bad news. The reason conception of understanding may fail all three of our tests. First, it does not adequately distinguish between knowing that and understanding why. In many cases (at least), to know that something is the case requires having reasons to believe it, so if the reason conception were correct, all these would also be cases of understanding why. But this is not so: there are many things we have reason to believe occur and know to occur, yet we do not understand why they occur. Given her expertise and honesty, the fact that your computer advisor tells you that your hard disk is severely fragmented gives you an excellent reason to believe that your hard disk is indeed severely fragmented; but it gives you not the slightest inkling *why* your disk is fragmented. Having a good reason to believe P is clearly not sufficient for understanding why P.

The reason conception is also under threat from the why regress. In one sense of reason, H does not provide a reason to believe E unless there is also a reason to believe H. On this construal of the notion of a reason, the reason conception of understanding would then entail that H can only explain E if H has itself been explained. What the why regress shows, however, is that H may explain E even if H is not itself explained. Finally, the reason conception does not readily account for self-evidencing explanations. If E is a reason for H, H cannot be a reason for E. If the spectral red shift is our reason for believing that the galaxy is receding, then the recession does not provide a reason for believing that the spectrum is shifted: this would be a vicious circle. So if the reason conception were correct, no self-evidencing explanations would be legitimate, but many are.

I turn now to my second contestant, the **familiarity** conception of understanding. This is the view that explanation is in some sense 'reduction to the familiar'. It is what is strange or surprising that we do not understand; a good explanation gives us understanding by making the phenomenon familiar, presumably by relating it to other things that are already familiar (cf. Hempel 1965, 430–3; Friedman 1974, 9–11). Loose though this specification is, it is enough to suggest that the familiarity conception of understanding, unlike the reason conception, may pass the first test. Something can be known yet also unfamiliar or surprising, so the familiarity conception leaves room for the gap between knowing that and understanding why. A further attraction of the familiarity view is the natural way it accounts for the fact that it is often surprise that prompts a request for explanation. It is often when things do not turn out as we expected that we want to know why. Moreover, even when we ask why about what is already in some sense familiar, the prompt for the question often involves 'defamiliarization': we are brought to see the everyday situation as somehow strange or surprising. The case of the moon already mentioned is a good example of this. Most people do not wonder why the same side of the moon always faces the earth, perhaps because they erroneously suppose that this is simply a consequence of the moon not spinning. Once they are shown that not only does the phenomenon require that the moon spin but also that its period be precisely the same as the apparently unrelated period of the moon's orbit around the earth, the phenomenon becomes surprising and prompts a why question.

The familiarity conception does not do as well on our other tests. It is unclear whether it allows for self-evidencing explanations. It is difficult to be sure about this without some specific and articulated account of what it takes to make a phenomenon familiar, but if H must itself be familiar in order to explain surprising E, it is unclear how E could provide an essential part of one's reason for believing H. It is odd to suppose that the surprising provides essential evidence for the familiar.

The familiarity conception also has difficulty with the why regress. If the conception entails that what is familiar is understood and that only what is familiar can explain, then it does not allow that what is not itself understood can nevertheless explain. But the why regress shows that we must allow for this: H may explain E even if we do not understand why H is the case.

The third view on our whirlwind tour is the **unification** conception of understanding. On this view, we come to understand a phenomenon when we see how it fits together with other phenomena into a unified whole (cf. Friedman 1974; Kitcher 1989). This conception chimes with the ancient idea that to understand the world is to see unity that underlies the apparent diversity of the phenomena. The unification conception allows for both the gap between knowledge and understanding and the legitimacy of self-evidencing explanations without difficulty. We can know that something is the case without yet being able to fit it together appropriately with other things we know, so there can be knowledge without understanding. Self-evidencing explanations are also accounted for, since a piece of a pattern may provide evidence for the pattern as a whole, while the description of the whole pattern places the piece in a unifying framework. The unification view may not do quite so well, however, on the why regress. Presumably a unifying explanation is itself unified, so there seems to be no room for explanations that we do not already understand. But this is not clear. For one might say that to explain a phenomenon is to embed it appropriately into a *wider* pattern. In this case H might suitably embed E, even though we have no wider pattern in which to embed H, and the requirements of the why regress would be satisfied.

Our fourth conception of understanding is that of **necessity**. The necessity view is that explanations somehow show that the phenomenon in question *had* to occur (cf. Glymour 1980). This conception of understanding acknowledges the gap between knowing that and understanding why, since one may know that something did in fact occur without knowing that it had to occur. The view also appears to allow for self-evidencing explanations, since there seems to be no vicious circularity involved in supposing that H shows E to be in some sense necessary while E gives a reason for believing H. It is less clear, however, that the necessity conception passes the why regress test: it fails the test if only what is itself necessary can confer necessity, or if only what is already known to be necessary can be used to show that something else is necessary too.

This leaves us with our fifth and final contestant, the **causal** conception of understanding. On this view, to explain something is to give information about its causes (cf. Lewis 1986; Humphreys 1989; Salmon 1998). The causal conception of understanding sails through our three tests. There is a gap between knowing and understanding, because we can know that something occurred without knowing what caused it to occur. The why regress is benign, because we can know that C caused E without knowing what caused C. Self-evidencing explanations are allowed, because it is possible for C to be a cause of E where knowledge of E is an essential part of one's reason for believing that C is indeed a cause.

The relative merits of the different conceptions of understanding are summarized in Table 1.1.

Because it does so well on our tests, and because so many explanations we give both in science and in everyday life are manifestly causal, the causal conception of understanding is my favourite, and will be my focus for the balance of this essay. But the causal conception is not without its difficulties (though I prefer the term 'challenges') and, in the spirit of full disclosure, I mention three of them here. The first is that we have no adequate account of causation; the second is that there are some explanations that seem clearly non-causal; the third is that not all causes are explanatory.

Table 1.1

TEST FEATURES	CONCEPTIONS OF UNDERSTANDING				
	Reason	Familiarity	Unification	Necessity	Causation
Knowledge versus Understanding	NO	YES	YES	YES	YES
Why Regress	NO	NO	MAYBE	NO	YES
Self-evidencing Explanation	NO	NO	YES	YES	YES

The problem of giving an account of the nature of causation is a hardy philosophical perennial. Most recent work is inspired, positively or negatively, by David Hume's enormously influential discussion (1777, Sec. VII). While many philosophers have offered solutions to the problem of the metaphysics of causation, none is generally accepted. (For a collection of recent work, see Sosa and Tooley 1993.) The second difficulty for the causal conception of understanding – the existence of non-causal explanations – is instantiated by mathematical and philosophical explanations, which are at least usually not causal. There also appear to be physical explanations that are noncausal. Suppose that a bunch of sticks are thrown into the air with a lot of spin so that they twirl and tumble as they fall. We freeze the scene as the sticks are in free fall and find that appreciably more of them are near the horizontal than near the vertical orientation. Why is this? The reason is that there are more ways for a stick to be near the horizontal than near the vertical. To see this, consider a single stick with a fixed midpoint position. There are many ways this stick could be horizontal (spin it around in the horizontal plane), but only two ways it could be vertical (up or down). This asymmetry remains for positions near horizontal and vertical, as you can see if you think about the full shell traced out by the stick as it takes all possible orientations. This is a beautiful explanation for the physical distribution of the sticks, but what is doing the explaining are broadly geometrical facts that cannot be causes.

The third and final difficulty for the causal conception is that not all causes are explanatory. Behind every event lies a long and dense causal history, most of which will not explain the event in a given context. When I ask my students why they have failed to hand in the supervision essays on time, I am unimpressed if they respond, 'Well, you know about the Big Bang ...'.

Nevertheless, I remain a fan of the causal conception of understanding. It is true that we have no adequate metaphysical understanding of causation, but as the why regress teaches us, this does not rule out the use of causal notions to illuminate other things. Nor in my view do we have a better grip on the central notions of any of the other four conceptions of understanding I have canvassed. As for the existence of non-causal explanations, this does show that causation cannot be the entire story of explanation. As I remarked above, the various conceptions of understanding are not mutually exclusive, so one can opt for more than one. Of the remaining four, the unification conception also did well on our tests, so this is another promising place to look; I also have some sympathy for the necessity conception. It seems clear, however, that very many of the explanations we give cite causes, and that in these cases what is said is explanatory precisely because what is cited is causal information. That leaves us with the difficulty that not all causes are explanatory. This really is in my view more a challenge than a difficulty, and one that we can go some way towards meeting. By giving a finer grained account of the context in which explanations are requested and of the why questions asked, we can give a causal account of explanation that itself explains why some causes are explanatory and others not. (For further discussion of recent work on explanation and understanding, see Salmon 1989 and Ruben 1993.)

Why do causes explain?

As we have seen, the test features of understanding support the causal conception. The gap between knowledge and understanding shows that the goods that explanations provide are more than the good provided by knowledge of the phenomenon to be explained. The why regress shows that the good of understanding is not like a substance that gets transferred from explanation to phenomenon explained, since H can provide an understanding of E even though we do not understand why H itself is the case. Self-evidencing explanations show that understanding does not involve providing a reason for belief. The causal conception respects these facts about understanding, and without portraying understanding as some mysterious form of super-knowledge, since although understanding E is more than knowledge of E, it need be no more than knowledge of the causes of E. Knowledge of causes is a primary good that many explanations provide.

In terms of philosophical explanations, the question we have been asking may be of the form 'Why is this a good explanation?', and the answer is 'It gives a cause, and causes explain.' This may be a good answer; but it is tempting to take another step up the why regress. Taking that step is to ask why causes explain. But does this question make sense? Or is it like asking why bachelors are unmarried? I think the question why causes explain does make sense, but it is difficult to articulate it in a way that makes this clear, and it is even more difficult to answer the question. I will struggle a bit with both these projects now.

In asking why causes explain, we are continuing our inquiry into the goods of explanation, but the question here does not simply concern the utility of causal knowledge. That question would be too easy. Knowledge of causes is useful for all sorts of reasons; but so is knowledge of effects. Yet while causes explain their effects, effects do not explain their causes. The recession of the galaxy explains why its light is red shifted, but the red shift does not explain why the galaxy is receding, even though the red shift may provide essential evidence of the recession. Part at least of the question I have in mind can be formulated contrastively. Why do causes rather than effects explain? Why don't effects explain their causes, given that causes explain their effects? These are more specific questions than the general question of why causes explain, but they are more than general enough for our purposes.

One may still feel that the question is silly. The reason causes explain and effects do not is simply that 'explanation' is a word we apply to causes and not to effects. But this does not do justice to the question. Our explanatory concepts and practices play an enormous role in cognitive economy, and one wants to know why this is the case. What is the point of this practice? This is just another way of asking about the goods of explanation, and to ask why we privilege causes over effects is a way of getting at part of this question.

Having made a pitch for the question of why causes explain rather than effects, I move briskly from the frying pan into the fire, because the question is very difficult to answer. In particular, it is difficult to avoid a more or less well-hidden dormative virtue explanation, along the lines of, 'causes explain because they, unlike effects, have the power to confer understanding'. Can we do any better than this? It is not clear. It is certainly not obvious that a thing's effects are any less important, useful, or interesting than its causes. And there is a clear sense in which finding out about a thing's effects increases our understanding of that thing. Indeed, one might argue that P's effects typically tell one more about P than do its causes. For effects often give information about P's properties in a way that causes do not. This is so because properties are at least often dispositional, and dispositions are characterized by their effects and not by their causes. Thus to say that arsenic is poisonous is to say roughly that if you eat it you will die. Thus the effects not only lead us back to the properties, but they are constitutive of at least some of them. In the conditional 'If you eat it, then you will die' there is both a cause and an effect, but they bear an asymmetrical relation to the corresponding property of being poisonous. Causing death is constitutive of the property of being poisonous, but eating arsenic, though a cause of death, is not constitutive of being poisonous. Nor do the causes of the arsenic or of its presence in a particular place appear to be constitutive of arsenic's properties.

A natural thought is that what is special about the causes of P is that they, unlike P's effects, create or bring about P. Can this be the key to the explanatory asymmetry between causes and effects? One worry is that this may be one of those dormative virtues stories, or worse. Why do causes explain effects? Because causes bring about effects. The worry is that 'bring about' is just another expression for 'cause', so all that has really been said is that causes explain because they are causes. One response would be to insist on a strong reading of 'bring about', a reading that would rule out a Humean account of causation, which takes causation to be no more than constant conjunction. Of course, Humeans may not like this, but they have the option of an error theory of explanation, according to which we never really explain why things happen, though the source of the illusion can be given, much as Hume himself had an error theory of necessary connection, according to which objects in the world are only conjoined, never connected, but the source of our mistaken idea of connection can be given (1777, Sec. VII). Such an error theory of explanation would treat understanding as a kind of myth, since it depends on a notion of causation that is metaphysically untenable. This would still be to allow that our notion of explanation and understanding, however misguided, depends on the idea of things being created by their causes. I would find such eliminativism about understanding unpalatable; but not being a committed Humean on matters causal, this line of argument does not overly concern me. Nevertheless, the thought that explanation depends on powerful metaphysical 'glue' linking E's cause to E strikes me as problematic for two other reasons. First, as one's account of causation strengthens the link between E's causes and E, it will do likewise for the connection between E and E's effects, so it is not clear that this appeal to a strong connection between cause and effect helps to explain the explanatory asymmetry that concerns us. Second, we often explain by appeal to causes that are not strongly connected to what they cause. This is well illustrated by explanatory causes that are omissions. A good answer to the question of why I am eating my campfire meal with a stick is that I forgot to pack my spoon, yet there seems no especially strong metaphysical link between the absence of the spoon and the use of the stick. Of course one may argue that explanations by omissions or negative causes are always oblique references to a positive causal scenario in which the process is strongly creative, but this strikes me as forced.

A closely related but I think better answer to our question of why causes rather than effects explain, though not without difficulties of its own, attributes the special explanatory power of causes to the link between causing and 'making a difference'. The idea is that causes explain because causes make the difference between the phenomenon occurring and its not occurring. This is connected to the idea of control, since we control effects through causes that make a difference, causes without which the effects would not occur. P's causes are handles which could in principle have been used to prevent P occurring in a way that P's effects could not. Of course control is not always an option. The galaxy's recession causes and explains its red shift even though we are in no position to change its motion; but the speed of recession is nevertheless a cause that made the difference between that amount of red shift and another. My suggestion is that this may partially explain why causes rather than effects yield understanding, since causes often make a difference in this sense while effects never do. Information about causes provide a special kind of intellectual handle on phenomena because the causes may make a difference and may themselves provide a special kind of physical handle on those phenomena.

I am far from confident that this difference view is correct, but I have four considerations that may count in its favour. First, given the obvious and enormous importance to us of knowledge of practical handles on phenomena, and the close link between control and making a difference, the difference view makes sense of our obsession about explanation. With all our leisure time, this interest has gone far beyond our practical concerns, but this overshooting is not particularly surprising. For one thing, given the difficulty of predicting which handles we will be able in time to pull, a broad strategy makes sense; for another we know that activities or traits originally caused by practical considerations may run way beyond the reasons for which they were originally selected, rather as an inclination to save potentially useful objects may lead to philately.

A second attraction of the difference view is that it may account for our ambivalence about the explanatory use of certain causes. For not all causes do make a difference. The obvious situation where they do not is one of overdetermination. A good ecological example is an environment with foxes and rabbits (Garfinkel 1981, 53–6). We ask why a rabbit is killed; we may answer by giving the location of the guilty fox shortly before the deed, or we may cite the high fox population. Both are causes but the details of the guilty fox's behaviour does not explain well because, given the high fox population, had that fox not killed the rabbit, another fox probably would have. Had the fox population been substantially lower, by contrast, the rabbit probably would have survived. The cause that made the difference is the cause that explains. This is some evidence for the difference view, though the situation is not entirely clearcut, since I think we often do judge the actual cause to have some explanatory power even when another cause would have done the job had the first one been absent. One possibility is that although a cause that made the difference is required (or strongly preferred) for explaining why, it is not required for explaining *how*.

The third consideration I adduce in favour of the difference view concerns contrastive explanation and brings out another way in which causes can fail to make the relevant difference and so fail to provide good explanations. Many of the why questions we ask are contrastive.

They have the form 'Why P *rather than Q?*', rather than simply 'Why P?', though the contrast often remains implicit, because it is obvious in the context in which the question is posed. Moreover, what counts as an explanatory cause depends not just on fact P but also on the foil Q. Thus the increase in temperature might be a good explanation of why the mercury in a thermometer rose rather than fell, but not a good explanation of why it rose rather than breaking the glass. We have already noted that not all of P's causes explain P in a given context; what we see now is that the foil in a contrastive question partially determines which causes are explanatory and which are not. And lo and behold, a good explanation requires a cause that made the difference between the fact and foil (Lipton 1993; 2004, ch. 3). Thus the fact that Smith had syphilis may explain why he rather than Jones contracted paresis (a form of partial paralysis), if Jones did not have syphilis; but it will not explain why Smith rather than Doe contracted paresis, if Doe also had syphilis. Contrastive explanations bring out the way in which what makes a difference between the P's occurring or not depends on what we mean by P not occurring, on our choice of foil. In so doing, it seems also to support the idea that the reason (some) causes explain is that they provide information about what made the salient difference between the occurrence and the non-occurrence of the effect of interest.

A final consideration that may support the difference view brings out a perplexing feature of explanation that I have not yet mentioned. This is the opacity of explanation, and it gives yet another way in which a cause (or a causal description) may fail to explain. For whether or not a cause explains depends on how it is described. This is clear, since one way of describing any cause of P is 'a cause of P', yet the question 'Why did P occur? is not illuminatingly answered by 'P occurred because of its causes'. To take a different example, suppose that the decayed insulation in the high-voltage lines running between the walls caused the fire in the department and is the event mentioned on page 17 of the accident report. If someone asks why the fire occurred, it is unhelpful to say 'Because of the event reported on page 17 of the accident report'. That oblique description does refer to a cause of the fire, but the description is not in itself explanatory (cf. Ruben 1990, 162–4).

It is not at all easy to say how we draw the demarcation between explanatory and unexplanatory descriptions of causes, but the idea of making a difference may help here too. The thought is that explanatory descriptions are those where changing the features described would make a difference. It is explanatory to say that the fire in the department occurred because of decayed insulation; it is not explanatory to say that the fire occurred because of the cause mentioned on page 17 of the accident report, however helpful that information may be in finding the explanation. Perhaps this is because, had the insulation not decayed, the fire would not have occurred, whereas it still would have occurred even if its causes where not mentioned in the report. In explanation we want a cause that makes the difference described in a way that tells us in virtue of what the difference is made.

An instrumental good of explanation

Having considered causal knowledge as one good that explanations deliver, and also the question of why causes should explain when effects do not, I end by flagging a quite different sort of good that explanations provide. In a word, this good is inference. This is an instrumental good, not part of understanding, but an example of how our explanatory practices are tools for the acquisition of other valuable things, in this case true beliefs. This is the idea behind Inference to the Best Explanation, a model designed to give a partial account of many inductive inferences, both in science and in ordinary life. Its governing idea is that explanatory considerations are a guide to inference, that scientists infer from the available evidence to the hypothesis which would, if correct, best explain that evidence. Many inferences are naturally described in this way. Darwin inferred the hypothesis of natural selection because, although it was not entailed by his biological evidence, natural selection would provide the best explanation of that evidence. To recycle my astronomical example, when an astronomer infers that a galaxy is receding from the earth with a specified velocity, she does this because the recession would be the best explanation of the observed red shift of the galaxy's spectrum. When a detective infers that it was Moriarty who committed the crime, he does so because this hypothesis would best explain the fingerprints, bloodstains, and other forensic evidence. In spite of what Sherlock Holmes would say, this is not a matter of deduction. The evidence will not entail that Moriarty is to blame, since it always remains possible that someone else was the perpetrator. Nevertheless, Holmes is right to make his inference, since Moriarty's guilt would provide a better explanation of the evidence than would anyone else's (cf. Lipton 2004).

Inference to the Best Explanation can be seen as an extension of one of the three test criteria that I used above to evaluate different notions of understanding. This is the prevalence of self-evidencing explanations, where the phenomenon that is explained in turn provides an essential part of the reason for believing the explanation is correct. According to Inference to the Best Explanation, this is a common situation in science: hypotheses are supported by the very observations they are supposed to explain. Moreover, Inference to the Best Explanation takes the idea of self-evidencing explanations one step further. It is not just that the observations support the hypothesis that explains them; it is precisely because that hypothesis would explain the observations that they support it.

Inference to the Best Explanation thus partially inverts an otherwise natural view of the relationship between inference and explanation. According to that natural view, inference is prior to explanation. First the scientist must decide which hypotheses to accept; then, when called upon to explain some observation, she will draw from her pool of accepted hypotheses. According to Inference to the Best Explanation, by contrast, it is only by asking how well various hypotheses would, if correct, explain the available evidence that she can determine which hypotheses merit acceptance. In this sense, Inference to the Best Explanation has it that explanation is prior to inference, and it is for this reason that inference can be a good that explanations deliver. This view complements the causal view of explanation nicely. Taken together, we have the idea that the construction and evaluation of competing explanations is one important route to the discovery of causes. If our explanatory practices give us this sort of information it is unsurprising that they play such a large role in our cognitive economy.

Conclusion

By asking about the goods of explanation, I have been seeking a kind of explanation of explanation. Philosophical explanations are perhaps particularly prone to the flaw of dormative virtues, where opium puts people to sleep because of its dormative powers. To say that we value explanations because they provide understanding is this sort of an inauspicious beginning. In the absence of an independent account of understanding, it gives us little more than the observation that we value explanations because of their explanatory power. This is the reason I began by considering different accounts of what understanding amounts to. Having settled on the causal view, I then considered the vexing question of why we should find causes explanatory and in particular why causes explain while effects do not, suggesting that only causes can make the relevant difference between the occurrence and non-occurrence of the thing we want explained. I then briefly suggested another explanatory good of a quite different order: not a type of understanding, but what understanding, especially causal understanding, is good for. It is good for causal inference.

The explanations of explanation I have sketched avoid dormative virtues, since the notions of causation, making a difference, and inference have the requisite independence from the notion of explanation itself. But what kind of explanations have I provided? Surprisingly perhaps, at least one of them is itself causal. I have suggested that one of the functions of explanation is inference. Functions are effects, but I go along with the view that function explanations are nevertheless causal, not 'effectal'. In the biological case, to explain the presence of a trait functionally is sometimes to use an effect as an oblique way of giving information about the evolutionary aetiology of the trait itself. Thus to say that the function of a polar bear's white fur is camouflage is to explain the presence of fur of that colour in terms of a causal history of evolution in which the possession of such fur by earlier bears or their progenitors offered a selective advantage and so caused there to be later bears with the same trait. To say that inference is a function of explanation may likewise be to provide a broadly causal explanation of the prevalence and persistence of our explanatory practices. Asking why things are as we find them to be provides us with an important way of discovering causes, and the fact that explanatory practices have this power is one of the reasons those practices have such a grip on us. It is curiously satisfying that we may thus give a causal explanation of causal explanation.

Acknowledgements

I am grateful to the Giora Hon, Sam Rakover, and Wesley Salmon for very helpful comments on an earlier draft of this paper.

References

Friedman, M. (1974). 'Explanation and scientific understanding'. *The Journal of Philosophy*, 71, 1–19.

Garfinkel, A. (1981). *Forms of explanation.* Yale University Press, New Haven.

Glymour, C. (1980). 'Explanations, tests, unity and necessity'. *Nous,* **14,** 31–50.

Hempel, C. (1965). *Aspects of scientific explanation.* Free Press, New York.

Hume, D. (1777). *An enquiry concerning human understanding.* (Ed. L. A. Selby-Bigg and P. H. Nidditch, Oxford University Press, 1975.)

Humphreys, P. (1989). *The chances of explanation.* Princeton University Press.

Kitcher, P. (1989). 'Explanatory unification and the causal structure of the world'. In *Scientific explanation,* xiii: *Minnesota studies in the philosophy of science* (ed. P. Kitcher and W. Salmon), pp. 410–505. University of Minnesota Press, Minneapolis.

Lewis, D. (1986). 'Causal explanation'. In his *Philosophical papers,* Vol. II, pp. 214–40. Oxford University Press, New York.

Lipton, P. (1993). 'Contrastive explanation'. In *Explaining explanation* (ed. D.-H. Ruben), pp. 207–27. Routledge, London.

Lipton, P. (2004). *Inference to the best explanation,* second edition. Routledge, London.

Ruben, D.-H. (ed.) (1993). *Explanation.* Oxford University Press.

Salmon, W. (1989). *Four decades of scientific explanation.* University of Minnesota Press, Minneapolis.

Salmon, W. (1998). *Causality and explanation.* Oxford University Press, New York.

Sosa, E. and Tooley, M. (ed.) (1993). *Causation.* Oxford University Press.

Steven Weinberg

2

Can science explain everything?
Can science explain anything?[1]

ONE evening a few years ago I was with some other faculty members at the University of Texas, telling a group of undergraduates about work in our respective disciplines. I outlined the great progress we physicists had made in explaining what was known experimentally about elementary particles and fields – how when I was a student I had to learn a large variety of miscellaneous facts about particles, forces, and symmetries; how in the decade from the mid-1960s to the mid-1970s all these odds and ends were explained in what is now called the Standard Model of elementary particles; how we learned that these miscellaneous facts about particles could be deduced mathematically from a few fairly simple principles; and how a great collective *Aha!* then went out from the community of physicists. After my remarks, a faculty colleague (not a particle physicist) commented 'Well, of course, you know science does not really explain things – it just describes them.' I had heard this remark before, but now it took me aback, because I had thought that we had been doing a pretty good job of explaining the observed properties of elementary particles and forces, not just of describing them.

I think that my colleague's remark may have come from a kind of positivistic angst that was widespread among philosophers of science in the period between the world wars. Ludwig Wittgenstein famously remarked that 'at the basis of the whole modern view of the world lies the illusion that the so-called laws of nature are the explanations of natural phenomena'. It might be supposed that something is explained when we find its cause, but an influential 1913 paper by Bertrand Russell[2] had argued that 'the word "cause" is so inextricably bound up with misleading associations as to make its complete extrusion from the philosophical vocabulary desirable.' This left philosophers like Wittgenstein with only one candidate for a distinction between explanation and description, one that is teleological, defining an explanation as a statement of the purpose of the thing explained.

E. M. Forster's novel *Where angels fear to tread* gives a good example of a teleological distinction between description and explanation. Philip is trying to find out why his friend Caroline helped to facilitate a marriage between Philip's sister and a young Italian man of whom Philip's family disapproves. After Caroline describes all the conversations she had with Philip's sister, Philip says, 'What you have given me is a description, not an explanation.' Everyone knows what Philip means by this – in asking for an explanation, he wants to learn Caroline's purposes. There is no purpose revealed in the laws of nature, and not knowing any other way of distinguishing description and explanation, Wittgenstein and my friend had concluded that these laws could not be explanations. Perhaps some of those who say that science describes but does not explain mean also to compare science unfavourably with theology, which they imagine to explain things by reference to some sort of divine purpose, a task declined by science.

This mode of reasoning seems to me wrong not only substantively, but also procedurally. It is not the job of philosophers or anyone else to dictate meanings of words different from the meanings in general use. Rather than argue that scientists are incorrect when they say, as they commonly do, that they are explaining things when they do their work, philosophers who care about the meaning of explanation in science should try to understand what it is that scientists are doing when they say they are explaining something. If I had to give an *a priori* definition of explanation in physics, I would say, 'Explanation in physics is what physicists have done when they say *Aha!*' But *a priori* definitions (including this one) are not much use.

As far as I can tell, this has become well understood by philosophers of science at least since the Second World War. There is a large modern literature on the nature of explanation, by philosophers like Peter Achinstein, Carl Hempel, Philip Kitcher, and Wesley Salmon. From what I have read in this literature, I gather that philosophers are now going about this the right way: they are trying to develop an answer to the question 'What is it that scientists do when they explain something?' by looking at what scientists are actually doing when they *say* they are explaining something.

Scientists who do pure rather than applied research commonly tell the public and funding agencies that their mission is the explanation of something or other, so the task of clarifying the nature of explanation can be pretty important to them, as well as to the philosophers. This task seems to me to be a bit easier in physics (and chemistry) than in other sciences, because philosophers of science have had trouble with the question of what is meant by an explanation of an event (note Wittgenstein's reference to 'natural phenomena'), while physicists are interested in the explanation of regularities, of physical principles, rather than of individual events. Biologists, meteorologists, historians, and so on are concerned not only with regularities, but also with individual events, such as the extinction of the dinosaurs, the blizzard of 1888, the French Revolution, etc., while a physicist only becomes interested in an event, like the fogging of Becquerel's photographic plates that in 1897 were left in the vicinity of a salt of uranium, when the event reveals a regularity of nature, such as the instability of the uranium atom. Philip Kitcher has tried to revive the idea that the way to explain an event is by reference to its cause, but which of the infinite number of things that could affect an event should be regarded as its cause?[3]

Within the limited context of physics, I think one can give an answer of sorts to the problem of distinguishing explanation from mere description, that captures what physicists mean when they say that they have explained some regularity. The answer is that *we explain a physical principle when we show that it can be deduced from a more fundamental physical principle*. Unfortunately, to paraphrase something that Mary McCarthy once said about a book by Lillian Hellmann, every word in this definition has a questionable meaning, including 'we' and 'a'. But here I will focus on the three words that I think present the greatest difficulties: 'fundamental', 'deduced', and 'principle'.

The troublesome word 'fundamental' can't be left out of this definition, because deduction itself doesn't carry a sense of direction; it often works both ways. The best example I know is provided by the relation between the laws of Newton and Kepler. Everyone knows that Newton discovered a law that says the force of gravity decreases with the inverse square of the distance, as well as a law of motion that tells how bodies move under the influence of any sort of force. Somewhat earlier, Kepler had described three laws of planetary motion: planets move on ellipses with the sun at the focus; a line from the sun to a planet sweeps over equal areas in equal times; and the squares of the periods (the times it takes the various planets to go around their orbits) are proportional to the cubes of the major diameters of the planets' orbits. It is usual to say that Newton's laws explain Kepler's. But historically Newton's law of gravitation was deduced from Kepler's laws of planetary motion. Edmond Halley, Christopher Wren, and Robert Hooke all used Kepler's relation between the squares of the periods and the cubes of the diameters (taking the orbits as circles) to deduce an inverse square law of gravitation, and then Newton extended the argument to elliptical orbits. Today, of course, when you study mechanics you learn to deduce Kepler's laws from Newton's laws, not vice versa. We have a deep sense that Newton's laws are more fundamental than Kepler's laws, and it is in that sense that

Newton's laws explain Kepler's laws rather than the other way around. But it's not easy to put a precise meaning to the idea that one physical principle is more fundamental than another.

It is tempting to say that more fundamental means more comprehensive. Perhaps the best known attempt to capture the meaning that scientists give to explanation was that of Carl Hempel. In his well-known 1948 article with Paul Oppenheim,[4] he remarked that 'The explanation of a general regularity consists in subsuming it under another more comprehensive regularity, under a more general law.' But this doesn't remove the difficulty. One might say, for instance, that Newton's laws govern not only the motions of planets but also the tides on earth, the falling of fruits, and so on, while Kepler's laws deal with the more limited context of planetary motions. But that isn't strictly true. Kepler's laws, to the extent that classical mechanics applies at all, also govern the motion of electrons around the nucleus, where gravity is irrelevant. So there is a sense in which Kepler's laws have a generality that Newton's laws don't have. Yet it would feel absurd to say that Kepler's laws explain Newton's, while everyone (except perhaps a philosophical purist) is comfortable with the statement that Newton's laws explain Kepler's.

This example of Newton's and Kepler's laws is a bit artificial, because there is no real doubt about which is the explanation of the other. In other cases the question of what explains what is more difficult, and more important. Here is an example. When quantum mechanics is applied to Einstein's general theory of relativity one finds that the energy and momentum in a gravitational field comes in bundles known as gravitons, particles that have zero mass, like the particle of light, the photon, but have a spin equal to two (that is, twice the spin of the photon). On the other hand, it has been shown that any particle of mass zero and spin two will behave just the way that gravitons do in general relativity, and that the exchange of these gravitons will produce just the gravitational effects that are predicted by general relativity. Further, it is a general prediction of string theory that there must exist particles of mass zero and spin two. So is the existence of the graviton explained by the general theory of relativity, or is the general theory of relativity explained by the existence of the graviton? We don't know. On the answer to this question hinges a choice of our vision of the future of physics – will it be based on space-time geometry, as in general relativity, or on some theory like string theory that predicts the existence of gravitons?

The idea of explanation as deduction also runs into trouble when we consider physical principles that seem to transcend the principles from which they have been deduced. This is especially true of thermodynamics, the science of heat and temperature and entropy. After the laws of thermodynamics had been formulated in the nineteenth century, Ludwig Boltzmann succeeded in deducing these laws from statistical mechanics, the physics of macroscopic samples of matter that are composed of large numbers of individual molecules. Boltzmann's explanation of thermodynamics in terms of statistical mechanics became widely accepted, even though it was resisted by Max Planck, Ernst Zermelo, and a few other physicists who held on to the older view of the laws of thermodynamics as free-standing physical principles, as fundamental as any others. But then the work of Jacob Bekenstein and Stephen Hawking in the twentieth century showed that thermodynamics also applies to large black holes, not because they are composed of many molecules but simply because they have a surface from which no particle or light ray can ever emerge. So thermodynamics seems to transcend the statistical mechanics of many-body systems from which it was originally deduced.

Nevertheless, I would argue that there is a sense in which the laws of thermodynamics are not as fundamental as the principles of general relativity or the Standard Model of elementary particles. It is important here to distinguish two different aspects of thermodynamics. On one hand, thermodynamics is a formal system that allows us to deduce interesting consequences from a few simple laws, wherever those laws apply. The laws apply to large black holes, they apply to steam boilers, and to many other systems. But they don't apply everywhere. Thermodynamics would have no meaning if applied to a single atom. To find out whether the laws of thermodynamics apply to a particular physical system, you have to ask whether the laws of thermodynamics can be deduced from what you know about that system. Sometimes they can, sometimes they can't. Thermodynamics itself is never the explanation of anything – you always have to ask why thermodynamics applies to whatever system you are studying, and you do this by deducing the laws of thermodynamics from whatever more fundamental principles happen to be relevant to that system.

In this respect, I don't see much difference between thermodynamics and Euclidean geometry. After all, Euclidean geometry applies in an incredible variety of contexts. If three people agree that each one will measure the angle between the lines of sight to the other two, and then they get together and add up those angles, the sum will be 180 degrees. And you will get the same 180 degree result for the sum of the angles of a triangle made of steel bars or of pencil lines on a piece of paper. So it may seem that geometry is more fundamental than optics or mechanics. But Euclidean geometry is a formal system of inference based on postulates that may or may not apply in a given situation. As we learned from Einstein's general theory of relativity, it does not apply in gravitational fields, though it is a very good approximation in the relatively weak gravitational field of the earth in which it was developed by Euclid. When we use Euclidean geometry to explain anything in nature we are tacitly relying on general relativity to explain why Euclidean geometry applies in the case in hand.

In talking about deduction, we run into another problem: Who is it that is doing the deducing? We often say that something is explained by something else without our actually being able to deduce it. For example, after the development of quantum mechanics in the mid-1920s, when it became possible to calculate for the first time in a clear and understandable way the spectrum of the hydrogen atom and the binding energy of hydrogen, many physicists immediately concluded that all of chemistry is explained by quantum mechanics and the principle of electrostatic attraction between electrons and atomic nuclei. Physicists like Paul Dirac proclaimed that now all of chemistry had become understood. But they had not yet succeeded in deducing the chemical properties of any molecules except the simplest hydrogen molecule. Physicists were sure that all these chemical properties were consequences of the laws of quantum mechanics as applied to nuclei and electrons. Experience has borne this out; we now can in fact deduce the properties of fairly complicated molecules – not molecules as complicated as proteins or DNA, but even some fairly impressive organic molecules – by doing complicated computer calculations using quantum mechanics and the principle of electrostatic attraction. Almost any physicist would say that chemistry is explained by quantum mechanics and the simple properties of electrons and atomic nuclei. But chemical phenomena will never be entirely explained in this way, and so chemistry persists as a separate discipline. Chemists do not call themselves physicists; they have different journals, they have different skills from physicists. It's difficult to deal with complicated molecules by the methods of quantum mechanics, but still we know that physics explains why chemicals are the way they are. The explanation is not in our books, it's not in our scientific articles, it's in nature; it is that the laws of physics require chemicals to behave the way they do.

Similar remarks apply to other areas of physical science. As part of the Standard Model, we have a well-verified theory of the strong nuclear force known as quantum chromodynamics, which we believe explains why the proton mass is what it is. It is not that we can actually calculate the proton mass; I'm not even sure we have a good algorithm for doing the calculation, but there is no sense of mystery about the mass of the proton. We feel we know why it is what it is, not in the sense that we have calculated it or even can calculate it, but in the sense that quantum chromodynamics can calculate it – the value of the proton mass is entailed by quantum chromodynamics, even though we don't know how to do the calculation.

Another problem with explanation as deduction: in some cases we can deduce something without explaining it. That may sound really peculiar, but consider the following little story. When physicists started to take the big bang cosmology seriously one of the things they did was to calculate the production of light elements in the first few minutes of the expanding universe. The way this was done was to write down all the equations that govern the rates at which various nuclear reactions took place. The rate of change of the abundance of any one nuclear species is equal to a sum of terms, each term proportional to the abundances of other nuclear species. In this way you develop a large set of linked differential equations, and then you put them on a computer that produces a numerical solution. When these equations were solved by Jim Peebles, and then by Bob Wagoner, Willy Fowler, and Fred Hoyle, it was found that after the first few minutes one quarter of the mass of the universe was left in the form of helium, and almost all the rest hydrogen, with other elements present in only tiny quantities. These calculations revealed certain regularities. For instance, if you put something in the theory to speed up the expansion, as for instance by adding additional species of neutrinos, you would find that more helium would be produced. This was somewhat counterintuitive – you might think speeding up the expansion of the universe would leave less time for the nuclear reactions that produce helium, but in fact the calculations showed that it increased the amount of helium produced.

The explanation is not difficult, though it can't easily be seen in the computer printout. While the universe was expanding and cooling in the first few minutes, nuclear reactions were occurring that built up complex nuclei from the primordial protons and neutrons, but because the density of matter was relatively low these reactions could occur only sequentially, first by adding protons and neutrons to make the nucleus of heavy hydrogen, the deuteron, and then by combining deuterons with protons or neutrons or other deuterons to make heavier nuclei like helium. However, deuterons are very fragile; they're relatively weakly bound, so essentially no deuterons were produced until the temperature had dropped to about a billion degrees, at the end of the first three minutes. During all that time neutrons were turning into protons, because neutrons are heavier than protons and so they can release energy by changing into protons. When the temperature dropped to a billion degrees and it became cold enough for deuterons to hold together, all the neutrons that were still left were rapidly gobbled up into deuterium, and the deuterium then into helium, a particularly stable nucleus. It takes two neutrons as well as two protons to make a helium nucleus, so the number of helium nuclei produced at that time was just half the number of remaining neutrons. So the crucial thing that determines the amount of helium produced in the early universe is how many of the neutrons have decayed before the temperature dropped to a billion degrees. The faster the expansion went, the earlier the temperature dropped to a billion degrees, so the less time the neutrons had to decay, so the more neutrons were left, and so the more helium was produced. That's the explanation of what was found in the computer calculations, but the explanation was not to be found in the computer-generated graphs of helium abundance vs. the speed of expansion.

Further, although I have said that physicists are only interested in explaining general principles, it is not so clear what is a principle and what is a mere accident. Sometimes what we think is a fundamental law of nature is just an accident. Kepler again provides an example. He is known today chiefly for his famous three laws of planetary motion, but when he was a young man he tried also to explain the diameters of the orbits of the planets by a complicated geometric construction involving regular polyhedra. Today we smile at this because we know that the distances of the planets from the sun reflect accidents in the way that the solar system happened to be formed. We wouldn't try to explain the diameters of the planetary orbits by deducing them from some fundamental law.

In a sense, however, there is a kind of approximate statistical explanation for the distance of the earth from the sun.[5] If you ask why the earth is about a hundred million miles from the sun, as opposed, say, to two hundred million or fifty million miles, or even further, or even closer, one answer would be that if the earth were much closer to the sun then it would be too hot for us and if it was any further from the sun then it would be too cold for us. As it stands, that's a pretty silly explanation, because we know that there was no advance knowledge of human beings in the formation of the solar system. But there is a sense in which that explanation is not so silly, because there are countless planets in the universe, so that even if only a tiny fraction are the right distance from their star and have the right mass and chemical composition and so on to allow life to evolve, it should be no surprise that creatures that inquire into the distance of their planet from its star would find that they live on one of the planets in this tiny fraction.

This kind of explanation is known as anthropic, and as you can see it does not offer a terribly useful insight into the physics of the solar system. But anthropic arguments may become very important when applied to what we usually call the universe. It is increasingly speculated by cosmologists that, just as the earth is just one of many planets, so also our big bang, the great expansion of the universe in which we live, may be just one of many bangs that go off sporadically here and there in a much larger mega-universe. It is further speculated that in these many different big bangs some of the supposed constants of nature take different values, and perhaps even some of what we now call the laws of nature take different forms. In this case, the question of why the laws of nature that we discover and the constants of nature that we measure are what they are would have a rough teleological explanation, that it is only in this sort of big bang that there would be anyone to ask the question. I certainly hope that what we now call the laws of nature and the constants of nature are accidental features of the big bang in which we happen to find ourselves, though constrained (as is the distance of the earth from the sun) by the requirement that they have to be in a range that allows the appearance of beings that can ask why they are what they are.

Conversely, it is also possible that a class of phenomena may be regarded as mere accidents when in fact they are manifestations of fundamental physical principles. I think this may be the answer to a historical question that has puzzled me for many years. Why was Aristotle (and many other natural philosophers, notably Descartes) satisfied with a theory of motion that did not provide any way of predicting where a projectile or other falling body would be at any moment during its flight, a prediction of the sort that Newton's laws do provide. According to Aristotle, substances tend to move to their natural positions – the natural position of earth is downward, the natural position of fire is upward, and water and air are naturally somewhere in between, but Aristotle did not try to say how fast a bit of earth drops downward or a spark flies upward. I am not asking why Aristotle had not discovered Newton's laws – obviously some one had to be the first to discover these laws, and the prize happened to go to Newton. What puzzles me is why Aristotle expressed no dissatisfaction that he had not learned how to calculate *when* objects will come to rest. This may have reflected a widespread disdain for change on the part of the Hellenic philosophers, as shown for instance in the work of Parmenides, which was admired by Aristotle's teacher, Plato. Of course Aristotle was wrong about this, but if you imagine yourself in his times, you can see how far from obvious it would have been that motion is governed by precise mathematical rules that might be discovered. As far as I know, this was not understood until Galileo began to measure how long it took balls to roll various distances down an inclined plane. It is one of the great tasks of science to learn what are accidents and what are principles, and about this we cannot always know in advance.

So now that I have deconstructed the words 'fundamental', deduce', and 'principle', is anything left of my proposal, that in physics we say that we explain a principle when we deduce it from a more fundamental principle? Yes, I think there is, but only within a historical context, a vision of the future of science. We have been steadily moving toward a satisfying picture of the world. We hope that in the future we will have achieved an understanding of all the regularities that we see in nature, based on a few simple principles, Laws of Nature, from which all other regularities can be deduced. These laws will be the explanation of whatever principles (such as, for instance, the rules of the Standard Model or of general relativity) can be deduced directly from them, and those directly deduced principles will be the explanation of whatever principles can be deduced from them and so on. Only when we have this final theory will we know for sure what is a principle and what an accident, what facts about nature are entailed by what principles, and which are the fundamental principles and which are the less fundamental principles that they explain.

I have now done the best I can to answer the second question in my title, so let me turn to the first: Can science explain everything? Clearly not. There will certainly always be accidents that no one will ever explain. There are questions like why the genetic code is precisely what it is or why a comet happened to hit the earth sixty-five million years ago in just the place it did rather than somewhere else that will probably remain forever outside our grasp. We cannot explain, for example, why John Wilkes Booth's bullet killed Lincoln while Puerto Rican nationalists who tried to shoot Truman did not succeed. We might have a partial explanation if we had evidence that one of the gunmen's arms was jostled just as he pulled the trigger, but as it happens, we don't. All such information is lost in the mists of time; it depends on accidents that we can never recover. We can perhaps try to explain them statistically: for example you might consider a theory that nineteenth-century Southern actors tended to be good shots whereas Puerto Rican nationalists tend to be bad shots, but when you only have a few singular events it's very difficult to make even statistical inferences. Physicists try to explain just those things that are not dependent on accidents, but in the real world most of what we try to understand does depend on accidents.

Further, science can never explain any moral principle. There seems to be an unbridgeable gap between 'is' questions and 'ought' questions. We can perhaps explain why people think they should do things, or why the human race has evolved to feel that certain things should be done and other things should not, but it remains open to us to transcend these biologically based moral rules. It may be, for example, that our species has evolved in such a way that men and women play different roles – men hunt and fight, while women give birth and care for children – but we can try to work toward a society in which every sort of work is as open to women as it is to men. The moral postulates that tell us whether we should or should not do so cannot be deduced from our scientific knowledge.

There are also limitations on the certainty of our explanations. I don't think we'll ever be certain about any of them. Just as there are deep mathematical theorems that show the impossibility of proving that arithmetic is consistent, it seems likely that we will never be able to prove that the most fundamental laws of nature are mathematically consistent. Well, that doesn't worry me, because even if we knew that the laws of nature are mathematically consistent, we still wouldn't be certain that they are true. You give up worrying about a certainty when you make that turn in your career that makes you a physicist rather than a mathematician.

Finally, it seems clear that we will never be able to explain our most fundamental scientific principles. (Maybe this is why some people say that science does not provide explanations, but by this reasoning nothing else does either.) I think that in the end we will come to a set of simple universal laws of nature, which we cannot explain. The only kind of explanation I can imagine (if we are not just going to find a deeper set of laws, which would then just push the question farther back) would be to show that mathematical consistency requires these laws. But this is clearly impossible, because we can already imagine sets of laws of nature that as far as we can tell are completely consistent mathematically but that do not describe nature as we observe it. For example, if you take the Standard Model of elementary particles and just throw away everything except the strong nuclear forces and the particles on which they act, the quarks and the gluons, you are left with the theory known as quantum chromodynamics. It seems that quantum chromodynamics is mathematically self-consistent, but it describes an impoverished universe in which there are only nuclear particles – there are no atoms, there are no people. If you give up quantum mechanics and relativity then you can make up a huge variety of other logically consistent laws of nature, like Newton's laws describing a few particles endlessly orbiting each other in accordance with these laws, with nothing else in the universe, and nothing new ever happening. These are logically consistent theories but they are all impoverished. Perhaps our best hope for a final explanation is to discover a set of final laws of nature and show that this is the only logically consistent *rich* theory, rich enough, for example, to allow for the existence of us. This may happen in a century or so, and if it does then physicists will be at the extreme limits of their power of explanation.

Notes

1. Talk given in a symposium on 'Science and the limits of explanation' at Amherst College, 20 October 2000. First published in *The New York Review of Books*, 2001.

2. 'On the notion of cause', reprinted in *Mysticism and logic* (Doubleday, Garden City, NY, 1957).

3. There is an example of the difficulty of explaining events in terms of causes that is much cited by philosophers. Suppose it is discovered that the mayor has paresis. Is this explained by the fact that the mayor had an untreated case of syphilis some years

earlier? The trouble with this explanation is that most people with untreated syphilis do not in fact get paresis. If you could trace the sequence of events that led from the syphilis to the paresis, you would discover a great many other things that played an essential role – perhaps a spirochete wiggled one way rather than another way, perhaps the mayor also had some vitamin deficiency – who knows? And yet we feel that in a sense the mayor's syphilis is the explanation of his paresis. Perhaps this is because the syphilis is the most dramatic of the many causes that led to the effect, and it certainly is the one that would be most relevant politically.

4. C. Hempel and P. Oppenheim. 'Studies in the Logic of confirmation'. *Philosophy of Science* (1948) **18**, 135; reprinted with some changes in *Aspects of scientific explanation and other essays in the philosophy of science* (Free Press, New York, 1965).

5. Professor R. J. Hankinson of the University of Texas tells me that this 'explanation' was given by Galen.

Martin Rees

3

Explaining the universe

Introduction

In August 1999, a solar total eclipse was visible from England – the first such event since 1927. I viewed it from Cornwall, through intermittent cloud. It was an 'environmental' experience shared with thousands of new age cultists, astrologers, and the like – I wasn't carrying out any scientific measurements. But it set me thinking about the scope and limits of science, and particularly the distinction between prediction, explanation, and understanding.

Eclipses could be predicted (at least approximately) even in ancient times. Throughout most of the first millennium BC, the Babylonians systematically recorded celestial events on cuneiform tablets, thousands of which are now in the British Museum. These records stretched over a long enough timespan to reveal subtle patterns – particularly the 18-year Saros cycle – which could be extrapolated forward to yield the dates when future eclipses were likely. This feat required no insight into how the sun and moon actually moved – only a faith in the regularity of nature.

There was no great advance on Babylonian predictions until the seventeenth century. Edmond Halley, one of the most versatile scientific polymaths of his time, is best known for the comet he studied in

1682: he predicted its return 76 years later, but didn't himself survive to see it again. However, he had the good luck that, during his lifetime, two total eclipses were visible from southern England, in 1715 and 1724. By that time, astronomers understood the layout of the solar system, and the intricacies of the moon's orbit – the 18-year period was realized to be due to a 'wobble' in the plane of the moon's orbit. Halley's predictions of the time, and of the location of the band of totality, surpassed the precision that the ancients could have achieved. But more important than this quantitative improvement was a qualitative advance: Halley, unlike the Babylonians, had the kind of insight that we could properly call an explanation.

Eclipses were predicted before they could be explained. We can now forecast them centuries ahead, to the nearest second; but those of us who journeyed to Cornwall to be in the predicted totality zone didn't know whether it would be clear or cloudy. We had only probabilities even just a few hours ahead. This is no reflection on the relative competence of astronomers and weather forecasters: it illustrates something crucial about the nature of explanation and understanding.

The earth's weather is far more complicated – harder to model – than the orbits of planets. But the difference is more fundamental than that. Weather patterns are *unstable*. In computing the earth's orbit, then insofar as our initial data are inaccurate, so will be our prediction. But the error takes a very long time to grow much larger than the uncertainty in the initial data. Our predictions of eclipses are nearly as accurate as our knowledge of the present positions and orbits of the earth and moon. On the other hand there are many systems where the uncertainty 'blows up' so that even a tiny difference in the starting point can make a big difference to the outcome – like a pencil balanced on its end. The way a butterfly flaps its wings today may determine the onset of storms a week later. That's why we can't make long-range weather forecasts, except of a statistical kind. Most physical phenomena are as seemingly capricious as the weather; the biosphere is still more vulnerable to time and chance. The 'clockwork' of the heavens is actually an atypical paradigm – the exception rather than the rule. Newton was lucky to have hit on one of the very few things in nature that are predictable. In general, we can't make predictions even when the underlying phenomenon obeys laws as deterministic as those that govern the motions of stars and planets.

The extinguishing of the sun, without forewarning, inspired dread in ancient times. Herodotus records such an event in 585 BC during a battle between the Medes and Lydians. The Babylonians learnt to predict eclipses, but without understanding. For the last three hundred years, eclipses are among the few things we can both understand and predict. Insofar as we do, we are not irrationally frightened of them. Science may have dispelled their mystery and terror, but that doesn't diminish their visual impact and wonder. As we develop more satisfying explanations we feel more at home in the universe. When the weather is violent, we can understand how it is the outcome of flow patterns in the atmosphere.

Explanations for the weather, though *post hoc*, are grounded in genuine understanding. But people crave explanations, even when there is not underlying understanding. A 'chill' is blamed on draughts, damp socks, or some other conjectured cause. Likewise, erratic stock market movements always find a ready explanation in the next day's financial columns: they are attributed to 'sentiment that pessimism about interest rate movements was exaggerated' or to 'the view that company x had been oversold'. Of course these are always *a posteriori*. Commentators could offer an equally ready explanation if a stock had moved the other way.

The hierarchy of the sciences

There are plainly limits on what we can even understand – it would be astonishing if human brains were 'matched' to all aspects of the external world. Most of nature – even the inanimate world – is neither explained nor understood. Nonetheless, we are almost all, in a sense, reductionists: we believe that any system, however large and complex, whether living or inanimate, is governed by Schrödinger's equation. In practice, we can't solve this equation for anything more complicated than a single molecule. The complexity of the solution for any macroscopic object depends on how many atoms are involved, and also on the intricacy of internal structure (for instance, a living cell is vastly more complex than a regular crystal made of the same number of atoms).

We can envisage the sciences in a 'hierarchy of complexity' – with atoms and crystals at the base, and (I suppose) psychology and sociology at the apex. But that, most emphatically, doesn't mean that all other sciences are 'applied physics'. Each of them, from chemistry to social psychology, has its own irreducible concepts: 'more is different', as P. W. Anderson averred in an oft-quoted article.

If you're studying how water flows, you lose the essence of the phenomenon when you analyse the fluid into atoms: experts in fluid mechanics seek explanations in terms of concepts like wetness, vorticity, and turbulence. Likewise, what goes on in a computer could be attributed to electrons moving in complicated circuits, but that misses the essence, the logic encoded in those signals. And, as Steven Pinker says, 'human behaviour makes the most sense when it is explained in terms of beliefs and decisions, not in terms of volts and grams'.

Even if we had a 'hyper-computer' that could solve Schrödinger's equation for a complex macroscopic system, the output wouldn't yield any real understanding or insight. Each science is autonomous, with its own concepts, and its appropriate type of explanation.

Problems in chemistry, biology, and the environmental and human sciences remain unsolved because scientists haven't elucidated the patterns, structures, and interconnections – not because we don't understand subatomic physics well enough. The scientific edifice isn't like a building whose superstructure is imperilled by an insecure base.

As a corollary, even if physicists achieved a 'final theory', the challenges of all the other sciences would remain as daunting as ever. Richard Feynman used a nice metaphor to make this point. Imagine you'd never seen chess being played before. Then, by watching a few games, you could infer the rules. Likewise, physicists learn the laws and transformations that govern the basic elements of nature. But in chess, learning how the pieces move is just a trivial preliminary on the absorbing progression from novice to grand master. Likewise, even if we know the basic laws, exploring how their consequences have unfolded over cosmic history is an unending quest. Our quest to understand the natural world is impeded far more by its *complexity* than by our ignorance of quantum gravity, sub-nuclear physics, and the like.

There is a sense – a limited one – in which some sciences can claim to be specially 'fundamental'. Causal chains – if you go on asking why? why? why? – lead back to a question in particle physics or cosmology. As Steven Weinberg has emphasized, it's an important feature of the universe that these arrows all point down and converge on a fundamental question either in particle physics or in cosmology – the sciences of the very small and the very large. These subjects deal with deep aspects of reality: we pursue them for that reason – not because the rest of science depends on them. As John Barrow (Chapter 5) reminds us, they are, to a mysterious degree, grounded in mathematics of a kind that human minds are attuned to.

The sciences are often likened to different levels of a building – logic in the basement, mathematics on the first floor, then particle physics, then the rest of physics and chemistry, and so forth as we climb upwards. But the analogy's poor. The superstructures – the 'higher-level' sciences dealing with complex systems – are autonomous, and aren't imperilled by an insecure base of 'emergent' concepts that only have meaning on scales large enough to allow us to treat a liquid as a continuum and ignore its atomic structure.

The subatomic world is simple. On the largest scales, our universe is simple too – that's why it isn't presumptuous to do cosmology. But almost everything in between is too complicated. Humans, the most complex entities we are yet aware of, are poised between cosmos and microworld. It would take about as many human bodies to make up the sun's mass as there are atoms in each of us.

There are 'laws of nature' in the macroscopic domain which are just as much of a challenge as anything in the microworld, and are conceptually autonomous from it – for instance, those that describe the transition between regular and chaotic behaviour, and which apply to phenomena as disparate as dripping taps and animal populations.

This issue became heated during the debates in the US about the Superconducting Super-Collider (SSC), a vast accelerator planned to be built in Texas at a cost exceeding 10 billion dollars. The hyperbole of its advocates led to a backlash from other scientists – even from some physicists – who felt that complex materials, fluid flows, and chaos theory offered intellectual challenges as great as those in particle physics.

Cosmology as a special science

As our understanding of fundamental science advances, it will probably reveal profound links between the 'inner space' of particles and the 'outer space' of the cosmos. My own professional interest is primarily in cosmology, the science of the very large, and perhaps also the grandest of the environmental sciences. But before delving more into this subject, let me offer a disclaimer. I'm often asked: Isn't it presumptuous to claim to explain *anything* about the vast cosmos? My response is that what makes things hard to understand is how *complicated* they are, not how big they are. A star is simpler than an insect, which embodies layer upon layer of structure. Biologists, tackling the intricacies of butterflies and brains, face tougher challenges than astronomers.

The range of modern telescopes extends out to distances of billions of light years. On these 'ultra broad-brush' scales a surprising simplicity and uniformity seems to prevail. Spectral studies of the light from the most distant galaxies reveal that they are made of atoms identical to those here on earth. Moreover, on the largest scales, our universe seems to be remarkably uniform and unstructured. Stars are assembled into the huge gravitating swarms we call galaxies; the galaxies are themselves grouped into clusters; and there are even superclusters. But we don't live in a fractal universe where there are clusters of clusters of clusters ... *ad infinitum*. Beyond a certain scale, the universe looks smooth. To take an analogy; if you are in the middle of the ocean, you are typically surrounded by an elaborate pattern of waves, small riding on large. But even the longest ocean swell is still small compared to the horizon distance. You can, in imagination, divide the visible ocean surface into separate patches, each large enough to be a fair sample: it makes sense to talk about the statistics, and the average properties, of the waves out to your horizon. This contrasts with, for instance, a mountain landscape, where a single peak can dominate the entire view.

The structures within our observable universe resemble an ocean surface rather than a mountain landscape – the largest conspicuous features are still far smaller than the horizon. We can meaningfully define quantities like the average density, the mean expansion rate, etc. Without this simplifying feature, scientific cosmology would not have got far.

(This ocean analogy prompts a further question. As far as the eye can see – out to the horizon a few miles away – the waves may be statistically the same. But conditions plainly aren't the same over the whole ocean: there are storms, and calms; and, even though the ocean may stretch far beyond the horizon – for thousands of miles, perhaps – it is eventually bounded by a shoreline. Could there, likewise, be very different domains far beyond our horizon? I shall return to this later.)

Thirteen billion years of cosmic history: from simple big bang to complex cosmos

Just as Darwinians and geologists have elucidated the 4.5 billion years of our earth's history, so cosmologists can set our entire solar system in a grand evolving scenario stretching back to a so-called 'hot big bang' – an era when everything was hotter than the centres of stars, and expanding on a timescale of a few seconds. It's beyond my brief to justify this claim in this paper: let me just say that the evidence is as compelling as most of the claims made by geologists and palae-ontologists about the history of the earth; indeed it's a lot more quantitative.

We can evolve 'virtual universes' in our computers, showing how gravity led to the emergence of structures, that developed into galaxies of stars. During the 13-billion-year expansion, our universe has evolved from a dense fireball into a cosmos with the intricate complexities we see around us and of which we're part. The first stars were made of just hydrogen and helium. But stars are fuelled by nuclear reactions, and thereby built up the periodic table from pristine hydrogen, cycling this processed material back into interstellar space via stellar winds or supernova explosions. Later generations of stars formed, each surrounded by dusty discs, which condensed into retinues of planets. One such star was our sun; one of its planets, the young earth, offered conditions propitious for the 'green fuse' of life to ignite. We and the earth are stardust – or less romantically, the nuclear waste from the fuel that makes stars shine.

That this elaborate intricacy all emerged from a boringly amorphous fireball might seem to violate a hallowed physical principle – the second law of thermodynamics, C. P. Snow's ancient touchstone of scientific literacy. This law describes an inexorable tendency towards uniformity, and away from patterns and structure: things tend to cool if they're hot, and to warm up if they're cold. Ink and water can readily mix; the reverse process – stirring a murky liquid until the dye concentrates into a black drop – would astonish us; ordered states get messed up, but not the reverse. In technical jargon, 'entropy' can never decrease. An apparent decrease locally is always outweighed by an entropy increase elsewhere: the classic example is a steam engine, where the ordered motion of a piston is always accompanied by waste heat.

We need to rethink our intuitions, however, when gravity comes into play. As the universe expands, patches marginally denser than average are braked more by gravity, and lag further and further behind: the density contrasts *grow*, until overdense systems stop expanding altogether and form galaxies. And when stars form within these galaxies, the contrasts in density and temperature grow ever larger. Odd though it seems, stars *heat up* when they *lose* energy. Suppose that the fuel supply in the sun's centre were switched off. Its surface would stay bright, because heat diffuses from the even hotter core. If nuclear fusion didn't regenerate this heat, the sun would gradually deflate as energy leaked away (within about 10 million years, as Lord Kelvin realized in the nineteenth century). But this deflation would actually make the core *hotter* than before: gravity pulls more strongly at shorter distances, and the central temperature has to rise in order to provide enough pressure to balance the greater force pressing down on it. Objects held together by gravity have a negative specific heat: once gravity gets a grip on a system, it inexorably contracts. Gravity has thereby transformed the early universe – homogeneous except for very slight irregularities in density – into our present structured cosmos, where temperatures range from the blazing surfaces of stars (and their even hotter centres) to the coldness of intergalactic space.

How scientific can cosmology be?

Our present-day cosmic habitat has emerged from simple beginnings in a 'big bang'. Moreover, this emergence is determined by the expansion rate, the content of the universe (atoms, radiation, and dark matter particles of other kinds, etc). It is also sensitive to a few basic constants of physics: the strength of gravity; the force that binds atomic nuclei (which permitted the alchemy inside stars that synthesized the atoms of the period table); and so forth. What fixed these numbers, and set our universe expanding? The answer lies, if anywhere, in the extreme conditions in the first tiny fraction of a second.

Cosmology is the study of an 'all embracing' event, and this perhaps requires cosmologists to view the laws themselves in a different perspective from other scientists. In most systems there are 'initial conditions', and then there are laws, described by differential equations, that govern how the system develops. Moreover, the laws are somehow autonomous – the system doesn't 'feed back' on those laws.

Newton didn't know what set up the planets, but once they were there his equations determined their future courses. Furthermore, those equations – Newton's laws of motion – were presumed to be universal, not determined by the solar system itself.

In cosmology, one can't assume this disjunction. Indeed one hopes to interrelate the laws of cosmos and microworld and, better still, to interrelate the universe and the laws themselves.

Superstrings and the M-theory

The two great 'pillars' of twentieth-century science are quantum mechanics, crucial in the microworld, and Einstein's theory of gravity, which does not incorporate quantum concepts. But we have no single framework that reconciles and unifies them. This lack doesn't impede terrestrial science, nor indeed the advance of astronomy, because most phenomena involve *either* quantum effects *or* gravity, but not both. Gravity is negligible in the microworld of atoms or molecules, where quantum effects are crucial; conversely, quantum uncertainty can be ignored in the celestial realm of planets, stars, and galaxies, where gravity holds sway. But right back at the beginning, everything we can now see was squeezed (according to some theories) into a region the size of a single atom, and quantum vibrations could shake the entire universe. To understand the first instants after the big bang or the space and time near the 'singularity' inside black holes demands a unification of quantum theory and gravity.

Several approaches are being followed, but there is no consensus yet about which is the right one. The most ambitious and encouraging approach seems to be superstring theory (or its still grander generalization, which is now called 'M-theory'). Its key idea is that the fundamental entities in our universe are not points, but tiny string loops, and that the various sub-nuclear particles are different modes of vibration of these strings. Moreover, these strings are vibrating not in our ordinary space (with 3 spatial dimensions, plus time) but in a space of 10 dimensions.

The extra dimensions are 'wound up' on a tiny scale, and don't manifest themselves to our 'coarse' observations – rather as a sheet of paper looks like a one-dimensional line when rolled very tightly, even though it is actually a two-dimensional surface. Each point in our ordinary space is really a 6-dimensional origami wrapped up on a sub-microscopic scale. There seem to be five ways this wrapping can happen; at a still deeper mathematical level these may be related structures embedded in an 11-dimensional space.

There is still, however, an unbridged gap between the intricate complexity of 10- or 11-dimensional space and anything that we can observe or measure. Moreover, string theorists are stymied by the mathematics. When Einstein developed general relativity, or Heisenberg his 'matrix mechanics' approach to quantum theory, the appropriate concepts and theorems were already 'on the shelf', having been developed by nineteenth-century mathematicians. But theorists may need twenty-first-century mathematics to understand, for instance, why a universe ends up with three 'expanded' spatial dimensions rather than some different number. The nature of our world, and the forces governing it, would depend on exactly how the extra dimensions 'wrapped up'. How does this come about and are there a lot of different ways in which it could happen?

Superstring loops may be 20 powers of ten smaller than atomic nuclei – inaccessible to direct experiment. Some theorists hope that the extra dimensions may manifest themselves in subtle ways that can be measured. But we may not be so lucky. How, then, can we check whether the theory is right? There are earlier precedents for theories being taken seriously, even without direct empirical support, if they have a resounding ring of truth about them. For example, many physicists were receptive to Einstein's general relativity even when evidence was sparse. Einstein was himself more impressed by his theory's elegance than by any experiments – he was, by all accounts, underwhelmed by news of the 1919 eclipse results.

Likewise Edward Witten, the leading guru of superstrings, has said 'good wrong ideas are extremely scarce, and good wrong ideas that even remotely rival the majesty of string theory have never been seen'.

One compellingly attractive feature of superstrings, suggesting that they indeed offer a valid route to unification of all physical forces, is that gravity seems an inbuilt consequence. Some hope that other features of the physical world that we do observe may 'pop out' of the theory. If it yielded insight into why the microworld is governed (as it is) by three forces, and why it is populated by particular classes of particles, we would be disposed to take its other predictions seriously even if they couldn't all be directly tested.

Another hope is that superstrings may offer new insights into the concepts of the quantum. Richard Feynman said that 'nobody really understands quantum mechanics'. It works marvellously: most scientists apply it almost unthinkingly. As John Polkinghorne put it 'The average quantum mechanic is not more philosophical than the average motor mechanic.' But it has its 'spooky' aspects, which many thinkers from Einstein onwards have found hard to stomach, and it's hard to believe that we've already attained the optimum perspective on it. We are at the stage the Babylonians were with eclipses and the calendar: useful predictions but no good 'explanation'.

The lure of the 'final theory' is very strong. Ambitious students want to tackle the number-one challenge. But an undue focus of talent in one highly theoretical area can be unhealthy, and is sure to be frustrating for all but a very few exceptionally talented (or lucky) individuals.

I advise my own graduate students to multiply the importance of their thesis problem by the (small) probability they'll solve it, and maximize that product. I remind them also of Peter Medawar's wise remark that 'No scientist is admired for failing to solve a problem beyond his competence. The most he can hope for is the kindly contempt earned by utopian politicians.'

Apparent fine tuning

In our universe, intricate complexity has unfolded from simple laws. But we notice something rather remarkable: this emergence depended rather sensitively on those laws, and on the way the universe was 'set up' – on the basic numbers that describe the fundamental forces, the expansion rate, and so forth.

Different choices of the numbers, and of the physical laws, would yield a boring or sterile universe. There is perhaps an analogy here with formulae in mathematics. Some can have rich implications. The Mandelbrot set, for instance, is an extraordinary pattern: tiny parts of it each reveal intricate and beautiful structures however much they are magnified. This seemingly infinite variety is encoded by a short algorithm. But you can readily write down other algorithms, superficially similar, that yield very dull patterns. Perhaps everything in nature is the outcome of a fundamental set of equations that can be written on a T-shirt. If so, when their consequences are 'played out' they lead, like Mandelbrot's algorithm, to immense complexity and variety. Different equations would yield still-born universes.

There are various ways of reacting to the apparent fine-tuning:

Happenstance or coincidence

One hard-headed response is that we couldn't exist if these numbers weren't adjusted in the appropriate 'special' way: we manifestly *are* here, so there's nothing to be surprised about. I think there is more to it than that. I'm impressed by a metaphor given by the Canadian philosopher John Leslie. Suppose you are facing a firing squad. Fifty marksmen take aim, but they all miss. If they hadn't all missed, you wouldn't have survived to ponder the matter. But you wouldn't just leave it at that – you'd still be baffled, and would seek some further reason for your good fortune.

Providence

Others adduce the 'tuning' of the numbers as evidence for a beneficent creator who formed the universe with the specific intention of producing us (or, less anthropocentrically, of permitting intricate complexities to unfold). This is the tradition of the eighteenth-century 'natural theologian' William Paley and other advocates of the so-called 'argument from design'. Paley's arguments pertained mainly to the biological world – the eye, the opposable thumb, and so forth. And they have fallen from favour even among theologians in the post-Darwinian climate. But variants of it are now espoused by, for instance, John Polkinghorne: he writes that the universe is 'not just "any old world", but it's special and finely tuned for life because it is the creation of a Creator who wills that it should be so.'

A special universe drawn from an ensemble, or 'multiverse'

But there is an alternative perspective to the 'providence' argument. The analogy of the watch and the watchmaker could be off the mark. Instead, the cosmos maybe has something in common with an 'off the peg' clothes shop. If the stock is large, we aren't surprised to find one suit that fits.

Perhaps, then, our 'big bang' wasn't the only one. An infinity of separate universes may have expanded, evolving differently, and perhaps ending up governed by different laws. Most would be sterile, or at least unpropitious for complex evolution, but we naturally find ourselves in one of the special subset where the requisite 'fine tuning' prevailed. This may not seem an 'economical' hypothesis – indeed nothing might seem more extravagant than invoking multiple universes – but it is a natural deduction from some (albeit speculative) theories, and opens up a new vision of our universe as just one 'atom' selected from an infinite multiverse.

I think the multiverse genuinely lies within the province of science. This is because, as I'll outline later, we can already map out how to put it on a more credible footing; more importantly we can envisage how the concept might be refuted. But first, I'd like to make a few comments aimed at those who may be disposed to dismiss any such discourse as 'metaphysics' (a damning put-down from a physicist's viewpoint!).

Epistemology of other universes

First, an issue of semantics. The proper definition of 'universe' is of course 'everything there is'. But the entity traditionally called 'the universe' – what astronomers study, or the aftermath of 'our big bang' – may be just one of a whole ensemble, each one maybe starting with its own big bang. Pedants might prefer to redefine the whole ensemble as 'the universe'. But it is less confusing to leave the term 'universe' for what is traditionally connoted, even though this then demands a new word, the 'multiverse', for the entire ensemble of 'universes'.

But does it make sense to envisage other 'universes' disjoint from our own? The 'multiverse' is, of course, a highly speculative concept.

However, the question 'Do other "universes" exist?' is one for scientists – it isn't just metaphysics. The following chain of reasoning may not be absolutely compelling, but should at least erase any prejudice that the concept is absurd.

We can envisage a succession of 'horizons', each taking us further than the last from our direct experience:

Galaxies beyond the range of present-day telescopes

There is a limit to how deep in space, how far back in time, and how close to black holes, our present-day instruments can probe. Obviously there is nothing fundamental about this limit: it is constrained by current technology, and enlarges year by year.

Astronomers now deploy telescopes with 10-metre mirrors to reveal galaxies so far away that their light has been travelling towards us for 10 billion years. And we detect weak microwaves that seem to pervade all space – an afterglow of the 'fireball' phase of our universe. We also have 'fossil' evidence, from studies of the chemical composition of earth and stars, about the history of our own Galaxy – that is, of past events close to our own world-line. (Strictly speaking, we only have a brief 'snapshot' of each galaxy: remoter ones are observed at an earlier phase in their history, because the light now reaching us set out earlier. But the uniformity and isotropy around us supports a 'cosmic Copernican hypothesis': we can then infer that all regions have evolved the same way and that, when we look at a patch of the universe several billion light years away, it resembles the way our own locality would have looked several billion years ago.)

Many more galaxies will undoubtedly be revealed in the coming decades by projected telescope arrays in space. We would obviously not demote them into non-existence simply because they haven't been seen yet. When ancient navigators speculated about what existed beyond the boundaries of the then-known world, or when we speculate now about what lies below the oceans of Jupiter's moons Callisto and Ganymede, we are speculating about something 'real' – we are asking a scientific question. Likewise, questions about remote parts of our universe are genuinely scientific, even though we must await better instruments to check our conjectures.

Galaxies unobservable – even in principle – until a remote cosmic future

Even if there were absolutely no technical constraints on the power of telescopes, our observations are still bounded by a 'horizon', set by the distance that a signal, moving at the speed of light, could have travelled since the big bang. This horizon demarcates the 'shell' around us on which the redshift would be infinite. There is nothing special about the galaxies on this horizon, any more than there is anything special about the circle that defines the horizon when you're in the middle of an ocean. On the ocean, you can see further by climbing up your ship's mast. But our cosmic horizon can't be extended *unless the universe changes*, so as to allow light to reach us from galaxies that are now beyond it.

If our universe were expanding at a uniform rate – neither accelerating nor slowing down – any particular galaxy would recede from us at unchanging speed, and would have just the same redshift billions of years from now (though by then it would of course have moved further away). But if the expansion were decelerating, then when the universe had become (say) twice as old as it is now, a given galaxy would be less than twice as far away, and would be less redshifted. The far-future 'horizon' would then encompass some galaxies that are now undetectable even in principle. The 'horizon' only grows perceptibly over the aeons of cosmic evolution. It is, to be sure, a practical impediment if we have to await a cosmic change taking billions of years, rather than just a few decades (maybe) of technical advance, before a prediction can be put to the test. But does that introduce a difference of principle? Surely it is still meaningful to talk about these faraway galaxies, and the far longer time before they can be observed is a merely quantitative difference, not one that changes their epistemological status.

Galaxies that emerged from 'our' big bang, but are unobservable in principle, ever

But what about galaxies that we can *never* see, however long we wait? These are a feature of cosmological models where the expansion *accelerates* rather than slows down: distances then grow faster than in linear proportion to the time, and the transmission of signals becomes harder, not easier, at later times. Acceleration occurs if gravity (which tends to brake the expansion) is overwhelmed by a repulsion of 'anti-gravity': some form of energy (with associated negative pressure) which is latent even in empty space, and which resembles the 'cosmological constant' lambda that Einstein introduced in 1917. Distant galaxies that accelerate away from us get more and more redshifted – their clocks, as viewed by us, seem to run slower and slower, and freeze at a definite instant so that though they never finally disappear we would see only a finite stretch of their future. It's the same if something falls into a black hole: from a vantage point safely outside the hole, we would see our infalling colleagues freeze at a particular time, even though they experience, beyond the horizon, a future that is unobservable to us. (That future is finite – they are engulfed in the singularity.)

It now seems that we may actually inhabit an accelerating universe. For the first few billion years gravity would have braked the expansion, yielding deceleration; but thereafter the density became so low that gravity was overwhelmed by a repulsive force. Our remote descendants can never learn what happens, in the far future, to the galaxies that we can now see: they will accelerate away from us, and appear frozen in time when their redshifts approach infinity; we will never see what happens to them afterwards.

Still more important, this type of universe contains galaxies that a given observer would *never* see. There would (as in a decelerating universe) be galaxies so far away that no signals from them have yet reached us. If deceleration continued, light from galaxies beyond our present horizon would eventually reach us. But if the cosmic repulsion has overwhelmed gravity, we are now accelerating away from remote galaxies at an ever-increasing rate, so if their light hasn't yet reached us, it never will. Such galaxies aren't merely unobservable in principle now – they would be *beyond our horizon forever*. But if a galaxy is now unobservable, it hardly seems to matter whether it remains unobservable forever, or if it only comes into view after a trillion years. (And I have argued, in the 'Galaxies unobservable' section above, that the latter should certainly count as 'real'.)

Galaxies in disjoint universes

The never-observable galaxies in the previous section would have emerged from the same homogeneous 'big bang' as us. But suppose that, instead of causally disjoint regions emerging from a single big bang (via an episode of inflation), we envisage separate big bangs. Are space-times completely disjoint from ours any less real than regions that never come within our horizon in what we'd traditionally call our own universe? Surely not – so the 'other universes' too should count as 'real' parts of our cosmos.

This step-by-step argument (unsympathetic readers might dub it a 'slippery slope' argument!) suggests that whether other universes exist is part of science. But it is of course speculative science. The next question is, can we put it on a firmer footing? What does it explain?

Many scenarios could lead to multiple universes. Andrei Line, Alex Vilenkin, and others have performed computer simulations depicting an 'eternal' inflationary phase where many universes sprout from separate big bangs into disjoint regions of space-time. Alan Guth and Lee Smolin have, from different viewpoints, suggested that a new universe could sprout inside a black hole, expanding into a new domain of space and time inaccessible to us. And Lisa Randall and Ramesh Sundrum suggest that other universes could exist, separated from us in an extra spatial dimension; these disjoint universes may interact gravitationally, or they may have no effect whatsoever on each other. In the hackneyed analogy where the surface of a balloon represents a two-dimensional universe embedded in our three-dimensional space, these other universes would be represented by the surfaces of other balloons: any bugs confined to one, and with no conception of a third dimension, would be unaware of their counterparts crawling around on another balloon. Other universes would be separate domains of space and time. We couldn't even meaningfully say whether they existed before, after, or alongside our own, because such concepts make sense only insofar as we can impose a single measure of time, ticking away in all the universes.

Guth and Edward Harrison have even conjectured that universes could be made in the laboratory, by imploding a lump of material to make a small black hole. Is our entire universe perhaps the outcome of some experiment in another universe? Smolin speculates that the daughter universe may be governed by laws that bear the imprint of those prevailing in its parent universe. If so, the theological arguments from design could be resuscitated in a novel guise, further blurring the boundary between the natural and supernatural phenomena.

Parallel universes are also invoked as a solution to some of the paradoxes of quantum mechanics, in the 'many worlds' theory, first advocated by Hugh Everitt and John Wheeler in the 1950s. This concept was prefigured by Stapledon, as one of the more sophisticated creations of his *Star Maker*:

> Whenever a creature was faced with several possible courses of action, it took them all, thereby creating many ... distinct histories of the cosmos. Since in every evolutionary sequence of the cosmos there were many creatures and each was constantly faced with many possible courses, and the combinations of all their courses were innumerable, an infinity of distinct universes exfoliated from every moment of every temporal sequence.

None of these scenarios has been simply dreamed up out of the air: each has a serious, albeit speculative, theoretical motivation. However, one of them, at most, can be correct. Quite possibly none is: there are alternative theories that would lead to just one universe.

Firming up any of these ideas will require a theory that consistently describes the extreme physics of ultra-high densities, how structures on extra dimensions are configured, etc. But consistency is not enough: there must be grounds for confidence that such a theory isn't a mere mathematical construct, but applies to external reality. We would develop such confidence if the theory accounted for things we *can* observe that are otherwise unexplained. At the moment, we have an excellent framework, called the Standard Model, that accounts for almost all subatomic phenomena that have been observed. But the formulae of the Standard Model involve numbers, about 18 altogether, which can't be derived from the theory but have to be inserted from experiment. Any theory that gave some insight into why there are particular families of particles, and the nature of the nuclear and electric forces, would acquire credibility; we would then be disposed to pay serious regard to other predictions it made, even if we couldn't directly test them. Some theories about 'extreme physics', when applied to the ultra-early universe, yield many universes that sprout from separate big bangs into disjoint regions of space-time. Andrei Linde's 'eternal inflation' is just one such possibility that cosmologists have addressed. The other universes would never be directly observable, even in principle (they would come under the 'Galaxies in disjoint universes' category above); we couldn't even meaningfully say whether they existed 'before', 'after', or 'alongside' our own. The physics of the first microsecond is still speculative: it doesn't have the same foothold in experiment as the theories that describe the later and cooler stages of the expanding universe, not the same level of observational support. However, if superstrings (or some other equally comprehensive theory) were 'battle tested' by convincingly explained things we *could* observe, then if it predicts multiple universes we should take them seriously too, just as we give credence to what our current theories predict about quarks inside atoms, or the regions shrouded inside black holes.

If other universes exist, theory may also offer clues to a further key question about them: How much variety do they display? 'Are the laws of physics unique?' is a less poetic paraphrase of Einstein's famous question, 'Did God have any choice in the creation of the world?'. Answering it is a key scientific challenge for the new century. If there were something uniquely self-consistent about the actual recipe, then any big bang would be a re-run of ours. But a far more interesting possibility (which is certainly tenable in our present state of ignorance of the underlying laws) is that *the underlying laws governing the entire multiverse may allow variety among the universes*. What we call the laws of nature – those that govern everything we observe – may in this grander perspective be *local bylaws*, consistent with some overarching theory governing the ensemble, but not uniquely fixed by that theory. Any things in our cosmic environment – for instance, the exact layout of the planets and asteroids in our solar system – are accidents of history. Likewise, the recipe for an entire universe may be arbitrary.

The same balance between chance and necessity arises in biology. Our basic development – from embryo to adult – is encoded in our genes, but many aspects of our development are moulded by our environment and experiences. And there are far simpler examples of the same dichotomy: snowflakes, for instance. The ubiquitous six-fold symmetry is a direct consequence of the properties and shape of water molecules. But their immense variety depends on their environment – on the fortuitous temperature and humidity changes during each flake's growth. If we had a fundamental theory, we would know which aspects of nature were direct consequences of the bedrock theory (just as the symmetrical template of snowflakes is due to the basic structure of a water molecule) and which are (like the distinctive pattern of a particular snowflake) the outcome of accidents. The accidental features could be imprinted during the cooling that follows the big bang, rather as a piece of red-hot iron becomes magnetized when it cools down, but with an alignment that may depend on chance factors. They could have other contingent causes, such as the influence of another universe separated from ours in a fourth spatial dimension.

Some features of our universe may be accidents, even if others are not – for instance, the microphysical laws may be standardized throughout the multiverse, but the numbers that determine the size and shape of the universe (and, perhaps, the number called lambda that measures the energy in space itself) might be different in each.

Some universes might be sterile because they collapse too soon, or are too small. The recipe for any 'interesting' universe must include at least some very large numbers: clearly not much could happen in a universe that was so constricted that it contained few particles. Every complicated object must contain a large number of atoms; to evolve in an elaborate way it must also persist for a long time – many, many times longer than a single atomic event. A related requirement, harder to quantify, is the presence of stable 'building blocks' that can combine in a variety of complicated ways, as the chemical elements can in our own universe.

At the moment, the view that the key numbers governing our universe are accidents of cosmic history is no more than a 'hunch'. But it could be firmed up by advances in our understanding of the underlying physics. More importantly for its standing as a genuinely scientific hypothesis, it is vulnerable to disproof.

Suppose, for instance, that the basic numbers describing our universe are an arbitrary outcome of how it cooled down – the outcome of bylaws in our cosmic patch – and take different values in other universes. Their values may not be typical of the entire multiverse: our universe must, as we have seen, have been special and atypical to permit our existence. But we would need to think again – and would need to call into question the 'ready-made clothes shop' explanation of the apparent fine tuning – if the numbers turned out to be *even more special* than our presence requires.

Consider, for example, the newly discovered repulsive force that accelerates cosmic expansion. Is its value an 'accident'? It has to be below a threshold to allow protogalaxies to pull themselves together by gravitational forces before gravity is overwhelmed by cosmical repulsion. An unduly fierce cosmic repulsion would prevent galaxies from forming.

Current indications are that the repulsive force is 10–20% of this threshold – implying (if all values for this force were equally probable) that we live in a universe that is between the 10th and 20th percentile of anthropically allowed universes. However, if (contrary to current indications) this force (Einstein's lambda) was thousands of times smaller than it needed to be merely to ensure that galaxy formation wasn't prevented, we might suspect that it was indeed zero for some fundamental reason. (Or that it had a discrete set of possible values, and all the others were well above the threshold.) By applying similar arguments to the other numbers, we could check whether our universe is typical of the subset that could harbour complex life. If so, the multiverse concept would be corroborated.

A Keplerian argument

The multiverse concept might seem arcane, even by cosmological standards, but it affects how we weigh the observational evidence in some current debates. Our universe doesn't seem to be quite as simple as it might have been. About 5% of its mass is in ordinary atoms; about 25% is in dark matter (probably a population of particles that survived from the very early universe contains atoms, and dark matter) and the remaining 70% is latent in empty space itself.

Some theorists have a strong prior preference for the simplest universe and are upset by these developments. It now looks as though a craving for such simplicity will be disappointed.

Perhaps we can draw a parallel with debates that occurred four hundred years ago. Kepler discovered that planets moved in ellipses, not circles. Galileo was upset by this. In his *Dialogues concerning the two chief systems of the world* he wrote 'For the maintenance of perfect order among the parts of the Universe, it is necessary to say that movable bodies are movable only circularly.'

To Galileo, circles seemed more beautiful; and they were simpler – they are specified just by one number, the radius, whereas an ellipse needs an extra number to define its shape (the 'eccentricity'). Newton later showed, however, that all elliptical orbits could be understood by a single unified theory of gravity. Had Galileo still been alive when *Principia* was published, Newton's insight would surely have joyfully reconciled him to ellipses.

The parallel is obvious. A universe with at least three very different ingredients may seem ugly and complicated. But maybe this is our limited vision. Our earth traces out just one ellipse from an infinity of possibilities, its orbit being constrained only by the requirement that it allows an environment conducive for evolution (not getting too close to the sun, nor too far away). Likewise, our universe may be just one of an ensemble of all possible universes, constrained only by the requirement that it allows our emergence. So I'm inclined to go easy with Occam's razor: a bias in favour of 'simple' cosmologies may be as short-sighted as was Galileo's infatuation with circles.

If there were indeed an ensemble of universes, then we would find ourselves in one of the small and atypical subsets governed by laws that permitted complex evolution. The seemingly 'designed' features of our universe shouldn't surprise us, any more than we are surprised at our particular location within our universe. We shouldn't take the Copernican 'principle of mediocrity' too far. We find ourselves on a planet with an atmosphere, orbiting at a particular distance from its parent star, even though this is really a very 'special' and atypical place. A randomly chosen location in space would be far from any star – indeed it would most likely be somewhere in an intergalactic void millions of light years from the nearest galaxy.

Concluding perspective

Elucidating the ultra-early universe, and firming up (or perhaps refuting) the concept of the multiverse, are challenges for the twenty-first century. These challenges look less daunting if we look back at what has been achieved during the twentieth century. A hundred years ago, it was a mystery why the stars were shining; we had no concept of anything beyond our Milky Way, which was assumed to be a static system. In contrast, our panorama now stretches out for more than 10 billion light years, and cosmic history can be traced back to within a fraction of a second of the 'beginning'. While observations gradually improve, theories advance in a zigzag. There is a sawtooth advance as fashions come and go; the gradient is nonetheless upwards.

This progress is possible only because of the contingency – in principle remarkable – that the basic physical laws are comprehensible, and apply not just on earth but in the remotest galaxies, and even in the first few seconds of our universe's expansion. Only in the first millisecond of cosmic expansion, and deep inside black holes, do we confront conditions where the basic physics remains unknown.

There are three great frontiers in science: the very big, the very small, and the very complex. Cosmology involves them all. Within a few years, the cosmic number should be as well-measured as the size and shape of the earth have been since the eighteenth century.

Theorists must elucidate the exotic physics of the very beginning, which entails a new synthesis between cosmos and microworld: it would be presumptuous of me to suggest how long this may take. Such a theory, if it comes, would signal the end of an intellectual quest that started with Newton, and continued through Maxwell, Einstein, and their successors. It would deepen our understanding of space, time, and the basic natural forces governing the everyday world, as well as elucidating the ultra-early universe and the centres of black holes.

But as well as being a 'fundamental' science, cosmology is also the grandest of the environmental sciences. It aims to understand how a simple fireball evolved, over 13 billion years, into the complex cosmic habitat we find around us – how, here on earth, and probably in many biospheres elsewhere, creatures evolve able to reflect on how they emerged. And this may be the greatest challenge of all.

A key conceptual innovation has been the realization that our universe may be vastly larger than the domain we can now (or, indeed, can ever) observe. What we've traditionally called our universe may be just an 'atom' in an ensemble – a multiverse punctuated by repeated big bangs, where the underlying physical laws permit diversity among the individual universes.

Anthropic explanations will then be the best we can hope for, and cosmology will in some respects resemble the science of evolutionary biology.

William C. Saslaw

4

Does physics rule the roost of scientific explanation?

PHYSICS differs from mathematics in which, to quote Bertrand Russell, 'we never know what we are talking about, nor whether what we are saying is true', and from cosmology in which we push our questions back to such great distances and early times that our observations become highly distorted and lose information. By contrast, physics generally works in the here and now. It tries to find mathematical relations among measurable phenomena and quantitative concepts. In this, physics is generally regarded as being extraordinarily successful, an archetype of science.

And now I'm going to let you into the great secret of physics' success: it asks relatively *simple* questions. Over the last three centuries, physicists have refined the technique of asking questions they know they can answer. Questions about well-defined phenomena that usually involve limited and specific forms of interaction. Of course, these questions may start off somewhat vaguely, but they have to congeal and become capable of experimental comparison in order to make significant progress. The artistry and imagination of physics consists in finding these questions and the appropriate methods for their solution. These methods involve dispensing with extraneous, irrelevant, and unimportant aspects of the problem. It's a bit like Michelangelo describing how one creates a sculpture by starting with a large block of marble and removing the unnecessary chips.

Not everyone will agree, at least initially, on which chips are unnecessary. There will be lots of discussion and controversy over which concepts, quantities, and parameters are most important for a satisfactory explanation. Then there will be more controversy, at least for important frontier questions, on how to best relate these concepts for an elegant explanation.

So what is an elegant explanation? And how do we know, as physicists, that we've found it? How do the controversies settle down, and most practising physicists agree that the problem is solved and that it's time to search for a new one? First I'll discuss this briefly in a rather abstract way, and then give two more concrete examples. One example from classical physics, introduced some of the basic methodology which survives to the present and enables some physicists to feel that their methodology provides the supreme model for scientific explanation. The second example, from a branch of modern physics only about a dozen years old, questions the supremacy of this approach to explanation.

Working physicists usually seek to solve specific problems; finding a suitable problem is often the most difficult part of research. Some obvious problems may be too complex to solve or too general, or beyond current techniques, resources, or abilities. Other problems, though solvable, may not be important, or may lead to ambiguous or irrelevant results. There is also a reservoir of repetitious, incremental problems suitable for grant proposals and research evaluations. Except for this last case, there is seldom any way of knowing the outcome without extensive analysis, although most good scientists eventually develop their intuition for finding good problems. The figure illustrates, very schematically, general working paths involved in reaching an explanation for a physical phenomenon.

Ideas, experiments, and sometimes simple observations generate problems. No one really knows how good ideas arise. Suddenly, or

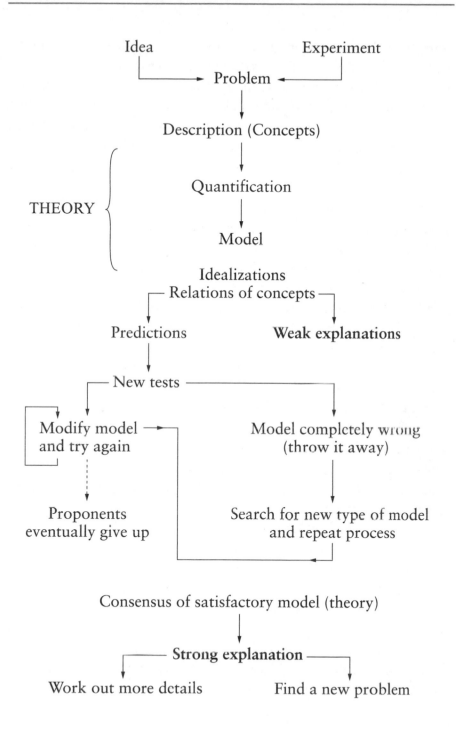

after long thought, we might be inspired with new relations among old concepts, or even with entirely new concepts. The best conditions to optimize scientific inspiration are largely unknown. They clearly vary for different people and different institutions, suggesting that it is very important to maintain diverse types of research environments. However, there is a very large set of commonly found conditions which inhibit and destroy inspirations. Inspiration needs to be nourished, for it is the beginning of all major progress in science.

Experiments sometimes uncover anomalies when pushed to greater precision, or they may reveal an entirely new phenomenon. Then we need to find if our previous scheme of things can explain our results, or if they require a new basis. From a rather abstract point of view, if we want to explain a physical phenomenon we first have to try to describe it. Not any description will do. It should be economical and precise. Economical in the sense that it omits extraneous properties. So if we want to explain why a table does not collapse under gravity – to take a homely example – we are not interested in its colour, or its history, or its surface texture. We are interested in its weight, its composition, its structure, and the strength of gravity. The starting point is to select the basic concepts on which to build our explanation. The precision of a description is clearly improved by mathematical quantification. This is not possible in all subjects, but it is a technique which physics has made great progress in developing over about the last three centuries.

Quantification in physics keeps us from going too far astray. If a description is purely qualitative and verbal, it is all too easy to bend the meanings of words and to modify conceptual relations to make a problem appear to be understood when it is not. As Lord Kelvin, one of the great physicists in the second half of the nineteenth century, put it:

> When you can measure what you are speaking about and express it in numbers, you know something about it; but when

> you cannot measure it, when you cannot express it in numbers, your knowledge is of a meagre and unsatisfactory kind: it may be the beginning of knowledge, but you have scarcely, in your thoughts, advanced to the stage of *science*.

Qualitative beginnings of knowledge often help provide significant insights in physics as well as in many other subjects. But combinations of quantitative and qualitative descriptions are more powerful than either alone. Calculation without understanding is fruitless; understanding without calculation is ambiguous.

Models result from the merger of understanding with calculation. These two components usually contribute with different weights depending on the type of problem and the style of the physicist creating or working out a model. It is not easy to define a model rigorously, though most working physicists know one when they see it. Models differ in style for different branches of physics, not to mention their differences between physics and other subjects. Very roughly, a model replaces a problem by a set of idealized relations among the concepts which describe the problem. Most models are specific to a problem, or a class of problems.

Some people call models theories, and others call theories models. This distinction does not seem to bother working scientists very much. They generally use the terms interchangeably, especially when distinguishing just between theory and experiment or observation. Reality, however, is more subtle: usually experiments and observations involve some interpretative theory, so they are never completely disentangled.

Models almost always involve major idealizations and simplifications of the actual problem. Like Michelangelo's sculpture, much is removed and the model remains. Understanding what to remove is a form of art, and involves some combination of inspiration, analogy, metaphor, and guesswork. What are the most important basic concepts? How should they be parameterized as physical quantities?

Could we do better with different concepts and fewer parameters? How are its concepts related to one another? Do any of them combine in especially interesting ways? And so on.

Many models have mechanical origins. Like toy engines, they are based on the rotating wheels, thrusting pistons, grinding gears, levers, and jiggling camshafts of everyday experience – such was the model that Maxwell originally used to explain his mathematical equations of electric and magnetic forces. Gradually, as Maxwell's models were developed they became more abstract. Forces were replaced by fields, fields by potentials, potentials by Lagrangians, and equations of motion by action integrals and modified metrics. All these were related to the phenomena one wished to explain.

These relations generally take two forms. If the model contains loose links and many parameters describing unclear concepts, it can, at best, only tie together stuff that is already known. These provide what might be called weak explanation. Economics, in trying to be quantitative, provides some familiar examples: fluctuation of stock markets, relations between growth and interest rates, inflation, and the balance of payments, etc. Models that only yield weak explanations tend to remain controversial and discourage consensus.

The other type of relation in models is so clear, tight, and well-defined that it can predict an observational or experimental result which has never been known before. Sometimes 'prediction' is used in a loose sense of a new (or even old) relation between quantities whose values are already known, but not directly incorporated into the model. Although this may have about the same logical status as a genuine prediction of the unknown, a real prediction seems much more psychologically satisfying and adds more conviction to the model. You get another psychological kick if the prediction is a surprise, coming from an unexpected relation. Naturally, all the relations in the model must also be consistent with one another.

Of course, for the model to carry this enhanced conviction, new tests of its predictions have to be arranged. If these new experiments or observations agree with the model, we may count ourselves fortunate – since it is historically unlikely – and look for more varied and more stringent tests of the model. The reason for this continued testing, rather than just celebrating the glories of a model that works, is that we never really understand a model until we know its boundaries, that is, under what conditions, how and why it breaks down and ceases to provide good explanations.

If the tests don't support the model, then we can either start afresh along quite different lines, or try modifying the model to get better agreement while maintaining the model's overall plausibility. At this stage, there are usually many different competing models and modifications. Then after a process of exhaustive examination and open criticism, more ideas and more experiments, the physicists working on the problem and repeating each other's experiments to check them often reach a consensus about the best model. This then provides what we may call a strong explanation: a generally agreed set of quantitative relations among well-chosen concepts which are consistent with many experiments or observations. This doesn't mean that the explanation is complete or completely compelling – just that it is relevant, robust (as checked by many modifications), and useful. A good strong explanation should set off an avalanche of understanding, not only of the original phenomenon to be explained but also of related problems. Even better, it may provide new analogies and inspiration for unrelated problems.

This is a rather general abstract view of how physicists actually develop explanations. An explanatory model of explanations if you will. Like the physical models it is also simplified and incomplete. For example, I have left out the social, historical, and psychological factors which enter at almost every stage of the development shown in the figure. But these are often specific to particular problems. So let me next give a brief glimpse into the development of explanations for two rather different types of physical problems – one ancient, the other recent.

Repercussions of the ancient problem ring throughout physics even today. It is, of course, the familiar problem of the orbits of our moon and the planets in our solar system. Newton thought on this for many years before being inspired – not by an apple but possibly by competition with Robert Hooke – to develop the mathematical model of a universal inverse square gravitational force propagating instantaneously through empty space and depending on the product of the masses of the interacting bodies. The planets, moon, and sun were idealized as particles, so none of their other properties mattered, at least at this level of approximation – just their masses, separations, and the value of the gravitational constant. It was a simple, elegant model, and Newton showed how it explained the known elliptical Keplerian orbits of the planets and predicted the detailed position of the moon.

Despite what some textbooks would have you believe, Newton's theory was not quickly accepted. Not quite as in Pope's epitaph 'Nature and Nature's laws lay hid in night: God said, let Newton be! And all was light.' Newton's model had a powerful rival by Descartes across the channel. Descartes' model supposed that all space was not empty, but filled with matter whose motion had condensed into vortices. Individual vortices dragged each planet in its orbit round the sun, and subsidiary vortices moved their moons within the larger vortices of the planets. Descartes recognized that the centrifugal force of circular motion would pull the planets away from their vortices. So to counteract this, he added an outer sphere of denser particles beyond the action of the vortices. The active pressure of these particles confined the planets.

Complicated, messy, and ad hoc? Yes! But Descartes' model had the advantage that it did not have to postulate a mysterious gravitational force strangely acting instantaneously at a distance. It was a rather common-sensical and satisfying everyday approach. After all, vortices had often been observed in water, and even occasionally in the air. And, as a bonus, it also explained sunspots as vortices on the solar surface.

In fact, Descartes' was the first mechanical theory of the solar system. Newton and others showed it was internally inconsistent, but Jean Bernoulli defended it, and it was seriously discussed as an explanation of the solar system in France until about 1744 when D'Alembert countered Bernoulli's arguments. This was a century after Descartes proposed his model, and decades after Newton had calculated the positions of the moon satisfactorily. Eventually Descartes' model died when it became clear that Newton's approach really did provide an avalanche of understanding for other phenomena in addition to celestial orbits.

Even more encouraging, Newton's model of forces between particles provided useful explanations or analogies for other very different phenomena. Rutherford's model of the atom was like a miniature solar system with the nucleus in the middle surrounded by orbiting electrons. Attracting electric forces between these particles held the atom together. Bohr improved on this by quantizing the orbits and Schrödinger abstracted the quantum orbits into wave functions. The transitions of electrons between these orbits released energy at definite frequencies which agreed with the observed radiation that atoms emit. Combining this with special relativity and Maxwell's electrodynamics gave the theory of quantum electrodynamics which has served as a model for modern particle physics. In turn, some particle physicists consider their models as providing insight into the earliest times of our universe and its inflationary expansion. So it all goes back to Newton, and has sprung out to apply his general style of explanation to the whole universe. Everything can ultimately be represented by the interactions of elementary particles. This is why some elementary particle physicists think they rule the roost of scientific explanation.

Others are not so sure. Although interacting particles may underlie all natural phenomena, some larger scale macroscopic physical processes are so complex that a microscopic particle explanation is essentially useless. Chemistry, solid state physics, and molecular biology, for example, are all subjects where one does not normally need to go below the atomic level for explanations of experiments. Even here, since these subjects often deal with the collective inter- actions of groups of particles, many of the concepts behind explan- ations have little to do with the ideas of particle physics, although all these subjects have exchanged useful analogies.

One of the useful modern concepts which, at least so far, has little to do with the particle–interaction mode of explanation is self- organized criticality. This is my second main example.

As a simple illustration of self-organized criticality, consider a pile of rice on a small table. Long-grained rice works best. (Originally sandpiles were thought to be good examples, but then experiments showed that the inertia of rolling sand grains modified their behaviour.) Now slowly add more rice to the top of the pile and watch what happens. Little avalanches. Lots of little avalanches, and sometimes big ones. It's a very interesting problem to ask how many avalanches there are of different sizes. Defining the size to be the num- ber of grains that fall together before the falling stops temporarily, it is a simple clear-cut problem. How do we solve it? Well, we know that each grain consists of many different rice molecules containing mostly hydrogen, oxygen, nitrogen, carbon, phosphorus, etc. atoms. They all interact in complicated ways, but I doubt if a model of these interactions will tell us how many rice grains will fall together. So let's think about it on the big scale of a whole grain. Each grain is an elongated, somewhat irregular, slightly compressible structure with a rather slippery surface. It is supported by an indeterminate number of other grains, irregularly disposed, and pushed down upon by a number of grains higher up in the pile, also irregularly disposed. The whole pile is in an essentially uniform gravitational field. Now I can add grains, one by one, to the centre of the pile, let the avalanches fall off the table, and ask for the distribution of avalanche sizes.

It seems an incredibly complicated question, even at this macro- scopic whole-grain level. We'd expect the results to differ for different size piles, to depend on what kind of rice we use, the history of how the pile was set up, how fast the rice falls, and perhaps some other properties. Yet when the experiment was done, by many physicists in many places, under controlled conditions, a remarkable result emerged. For a wide range of conditions, the number of avalanches $N(m)$ each containing m grains is a simple power law: $N(m) \propto m^{-a}$ with a ≈ 2. How can this be? What explains such a simple outcome of such a complicated situation?

The answer was found about a dozen years ago from a new type of model which doesn't take the complicated shapes and positions of the grains, or even the force of gravity directly into account at all. It is a cellular automaton. This is nothing more than a piece of graph paper divided into square cells, each containing some representative point particles, and several simple rules for moving the particles through the cells. The aim of this model is to extract the essence of the experiment. For the rice pile, first sprinkle the cells randomly with no more than three particles per cell. Then add a particle, either in the centre or to any random cell (it doesn't much matter). If the number of particles in a cell equals four, then redistribute all the four particles in that cell by putting one in each of the four neighbouring cells. Any particles that move off the grid at the edge are removed from the system. That's it. These are all the rules. The model can also be extended to three or more dimensions.

You might think that with such simple rules nothing very interesting would happen. And you would be right, up to a point. At first, only a rare cell would acquire four particles and topple onto its neighbours. But as you add more and more particles, there would be more cells with three particles, near the critical number of four particles. Then when an avalanche starts, it becomes more likely to propagate further through the surrounding cells. It's like a domino effect. When the number of particles becomes large enough, the system reaches a critical stage where avalanches of many sizes become unpredictable. This is a form of *self-organized* criticality. In this stage $N(m)$ satisfies a power law, similar to that of avalanches in rice piles. It is robust under minor changes of the rules, and does not require the external imposition of particular conditions on the system.

Thus this simple model of self-organized criticality captures an essential feature of the complex dynamics of rice pile avalanches. This model helped usher in a new subject in physics, complexity theory, about a dozen years ago. It has subsequently been extended, and tested in many ways, with both real experiments and computer model experiments. There is now a growing consensus among physicists that such self-organized criticality models provide a new type of explanation for some aspects of many phenomena, including rice piles, earthquakes, pendulums linked together, and fluctuations in some commodity markets. There have also been attempts to apply these and related ideas to explain features of the evolution of species, the origin of life, and even the way thoughts form in our brains. Whether the physical models of self-organized criticality can be transplanted to these other areas and provide useful intuition and explanations is still controversial and there is no consensus.

So, does physics rule the roost of scientific explanation? Many physicists would automatically answer, 'Yes. Of course! After all, we have had our Newton who showed the power of a really good explanation by actually constructing several of them.' And, they would go on to add, 'The physicist's style of explanation is frequently envied by other subjects where it is "often imitated but never excelled."'

However, we have seen that even physics is not monolithic. All its explanations do not rise from a base of elementary particles. Different types of phenomena need different types of explanations, as the rice pile illustrates. What physics does show is that if you ask a sufficiently simple, clear question, you can often get a clear insightful and convincingly quantitative answer. Formulating the question is not always easy, and the answer is never final. Techniques are available to help nearly everyone who has studied a problem in physics carefully to achieve a consensus of the best current explanation. Some of these techniques have been helpful in other subjects as well, often in ways not imagined initially. However, it is often counter-productive to try to apply them ritualistically or with insufficient insight. So we may conclude with a note by Beethoven who said 'In music everything should be both surprising and expected.' I think we can say the same about the muse of scientific explanation.

John D. Barrow

5

Mathematical explanation

Mathematics – a very peculiar practice

SCIENTISTS have long known that there is safety in numbers. Mathematics offers a sure path to understanding Nature's workings. This feature of the world – the remarkable effectiveness of simple mathematics for its description – was one of the first things that attracted me to sciences like astronomy as a teenager. It seems quite extraordinary that simple patterns deduced from watching gases expanding in jars, or the swinging of a bob on the end of a piece of string, could tell us things about the structure of stars or the expansion of the universe. Why was it that the next step the world took was the same as that predicted by squiggles on a piece of paper? Why does the universe march to a mathematical tune?

Mathematics is unlike all of the natural sciences because it is potentially infinite in extent. There is no end to the catalogue of mathematical structures that might be described or invented. Yet it is intimately linked to the process of explanation in all the sciences. If you can give a 'mathematical explanation' for something seen in Nature then that is pretty much sufficient for the consciences of most scientists.

The debate over what mathematics 'is' has been the subject of a host of books. Is it discovered? Is it invented? Is it logically constructed? Is it socially constructed? Is it emergent? While these questions are interesting they will not be the object of this discussion. Our only comment on them is the author's belief that it is quite wrong to imagine, as all the discussions of this question tend to argue, that *all* of mathematics falls into one category: that it is discovered in a 'pi in the sky' Platonic fashion *or* it is constructed. Even if some mathematics were divinely revealed to us by means of a glimpse into some Platonic heaven we would still be able to use that mathematics as the basis for other constructions.[1]

If you want to know what mathematics *is*, a mathematician is probably the last person to ask. Historians know what history is, sociologists know what sociology is, but most mathematicians neither know nor care what mathematics is: it is simply what mathematicians do. If you question your mathematical friends more closely, you will find a range of quite different views on offer: for some it is merely a game of logical patterns, like chess or chequers; for others it is the uncovering of the deep structure of reality, the nearest we get to thinking the unadulterated thoughts of God. In between these extremes there is plenty of room for compromises and variations.

When replying to the question 'What is mathematics?' to non-mathematicians it is helpful to categorize it, rather weakly, as the collection of all possible patterns. Some of those patterns are physically manifested; some are in sequences of numbers; other exist only in mental creations; while others are in logical inter-relations. When viewed in this way the ubiquity of mathematics as a description and an explanation for so much of the physical world is demystified. It is inevitable that the world is mathematical – that mathematics 'works'. For in any world that contains minds or observers that must be pattern of some sort and where there is pattern there is mathematics.

Yet, this perspective does not solve the puzzle of the 'unreasonable effectiveness of mathematics'. Why are such simple patterns so far-reaching in their explanatory power? It would be quite possible for the patterns of Nature to be described by mathematics but yet mathematics would not be very useful to use in practice because natural patterns were of a sort that did not readily admit simple algorithms for carrying out computations, or because even the simplest natural patterns were to our understanding as is quantum mechanics to Schrödinger's cat.

The utility of a mathematical explanation

The utility of mathematics is telling us that *computable* mathematical operations are common. A computable mathematical operation can be conducted by a finite sequence of step-by-step operations that could be implemented by a real or imaginary Turing machine that is able to change a string of digits into another string of digits. Although computer scientists are usually interested in discovering what aspects of the real world can be simulated by computers of this sort (the human mind in the case of AI), this idea can be inverted and the fact that computable operations can be mimicked by physical processes tells us that those physical processes are well described by certain mathematical functions. It would be possible for the world to be mathematical in a very deep sense but if the mathematics that characterized its patterns was non-constructive and non-computable then we would not find mathematics to be terribly useful for carrying out calculations, making predictions, or understanding what we see around us. Another way to characterize this utility of mathematics is to say that the world appears to be extremely 'compressible' in the algorithmic sense. This is, strings of symbols that we might use to encode information about the world around us can be replaced by much briefer rules, or formulae, or programs, which contain the same information. Thus if you were shown the sequence 0,2,4,6, 8,10,12,14, ... it could be replaced by the instruction to print the even numbers. You could convey the sequence to interested parties elsewhere by transmitting a formula shorter than the listing of the sequence. If a sequence cannot be condensed into a rule that is shorter than the listing of the sequence then we say that it is incompressible, or random. Science is the search for compressions in this sense. We can just go around the universe dressed in white coats gathering up lists of facts and figures, recording the positions of the planets at every instant of time. But this is not really doing science. Science begins when we try to compress the information content in those data strings by replacing them by formulae or rules from which they can be regenerated. More important, those same compressions can be used to predict what the data sequences will be found to be in other circumstances that have not yet arisen or have yet to be observed. These compressions are often called laws of Nature. The utility of simple mathematics in making sense of so much of the world around us is a reflection of the abundance of simple computable mathematical functions and the algorithmic compressibility of the universe.

The compressibility of the universe is linked to a number of features of its structure that make life simpler than it might have been for scientists. At the low energies where biochemical life is possible, there are only four known fundamental forces of Nature (weak, strong, electromagnetic, and gravitational). The fact that there are so few forces and they are so different in strength is important. Hundreds of forces would be fantastically complex to deal with and would require much larger symmetries and patterns to be employed to encode their behaviours into mathematics. The difference in the strengths of the forces ensures that they tend to come into play in pairs when equilibrium structures form. Many of the forces are also weak. This enables simple algorithms to be used to approximate their behaviour. Also important is the fact that these forces act upon families of identical elementary particles. In this way the nature of the particles governed by laws can be linked to the forms of the laws that control them. When we move up the scale from elementary particles to atoms and molecules it is the presence of quantization that guarantees the regularity and repeatability of the world. If energies were not quantified then every pair or an electron and a proton would be different. Even if they started with the same sizes they would drift away under the influence of the buffetings of other particles. Quantization ensures that hydrogen atoms are the same. It is the root of reproducibility in Nature.

One further mystery about mathematical explanation is its uncanny foresight. Time and again, physicists discover that the mathematical patterns needed to capture most completely the newly discovered properties of the universe turn out to be parts of 'pure' mathematics created long ago for no practical purpose. The more removed from everyday experience our mathematical explanations are, so the more accurate do they appear to be. The accuracy of our understanding of the spinning down of the binary pulsar or the properties of electrons far surpasses that of anything else that we know in the universe. If mathematics is entirely a human creation that is nothing more than a necessary way of thinking it is very strange that it becomes better and better as we go farther and farther away from the arena of local human experience and the realm in which the process of natural selection fashioned our mental categories.

Pure maths is bigger than applied maths

During the nineteenth century, mathematics began to move in a new direction and its scope expanded beyond the paths mapped out by the ancients. For them, mathematics provided a way of making precise statements about quantities, lines, angles, and points. It was divided into arithmetic, algebra, and geometry and formed a vital part of the ancient curriculum because it offered something that only theology would also dare to claim – a glimpse into the realm of absolute truth. The most important exemplar was geometry. It was the most impressive and powerful instrument wielded by mathematicians. Euclid created a beautiful framework of axioms and deductions that led to truths called 'theorems'. These truths led to new knowledge of the motions of the planets, new techniques for engineering and art; Newton's greatest insights were achieved by means of geometry.

Geometry was not seen as merely an approximation to the true nature of things, it was part of the absolute truth about the universe. Like part of some holy writ, the great theorems of Euclid were studied in their original language for thousands of years. They were true, perfectly so, and they provided human beings with a glimpse of absolute truths. God was many things but he was undoubtedly also a geometer.

We begin to see why mathematics was of such importance to theologians and philosophers. With no knowledge of mathematics you might have been persuaded that the search for absolute truth was a hopeless quest. How could we fathom its bottomless complexity given the approximate and incomplete nature of our understanding of everything else in the world around us? How could a theologian claim to know anything about the nature of God or the nature of the universe in the way that medieval philosophers seemed to do so confidently in their pronouncements about the vacuum and the void? Their justification was in the success of Euclid's geometry. It was the prime example of our success in understanding a part of the ultimate truth of things. And if we could succeed there, why not elsewhere as well? Euclid's geometry was not just a mathematician's game, a rough approximation to things, or a piece of 'pure' mathematics devoid of contact with reality. It was the way the world was. A similar exalted status was afforded the system of logic that Aristotle introduced as the means by which the truth or falsity of deductions made from premises could be ascertained. Aristotle's logic was accepted as being true and perfectly representative of the working of the human mind. It was the one and only way of reasoning infallibly.[2]

Later mathematical discoveries by Gauss, Bolyai, and Lobachevskii revealed Euclidean geometry to be but one of many possible logically self-consistent systems of geometry. All but one of these possibilities were *non-Euclidean*. None had the status of absolute truth. Each was appropriate for describing measurements on a different type of surface, which may or may not exist in reality. With this, the philosophical status of Euclidean geometry was undermined. It could no longer be exhibited as an example of our grasp of absolute truth.

From this discovery would spring a variety of forms of relativism about our understanding of the world.[3] There would be talk of non-Euclidean models of government, of economics, and of anthropology. 'Non-Euclidean' became a by-word for non-absolute knowledge. It also served to illustrate most vividly the gap between mathematics and the natural world. Mathematics was much bigger than physical reality. There were mathematical systems that described aspects of Nature, but there were others that did not. Later, mathematicians would use these discoveries about geometry to discover that there were other logics as well. Aristotle's system was, like Euclid's, just one of many possibilities. Even the concept of truth was not absolute. What is false in one logical system can be true in another. In Euclid's geometry of flat surfaces, parallel lines never meet, but on curved surfaces they can.

These discoveries revealed the difference between mathematics and science. Mathematics was something bigger than science, requiring only self-consistency to be valid. It contained all possible patterns of logic. Some of those patterns were followed by parts of Nature; others were not. Mathematics was open-ended, uncompleteable, infinite; the physical universe was smaller.

The proliferation of mathematical systems led to the notion of what is now called 'mathematical modelling'. Particular pieces of mathematics help us describe aerodynamic motion but if we want to understand risk and chance we may have to use other mathematics. On the purer side of mathematics, it was recognized that there exist different mathematical structures, each defined by the objects (for

example numbers, angles, or shapes) they contain and the rules for
their manipulation (like addition or multiplications). These structures
have different names according to the richness of the rules that are
allowed.

These simple systems of mathematics they had abstracted from
the natural world provided models from which new abstract structures, defined solely by the rules for combining their symbols, could
be created. Mathematics was potentially infinite. The subset of mathematics which described parts of the physical universe was smaller,
perhaps even finite. Each mathematical structure was logically independent of the others. Bertrand Russell, writing in 1901, captured its
new spirit as eloquently as anyone:

> Pure mathematics consists entirely of such asseverations as
> that, if such and such a proposition is not true of *anything*, then
> such and such another proposition is true of that thing. It is
> essential not to discuss whether the first proposition is really
> true, and not to mention what the anything is of which it is sup-
> posed to be true ... If our hypothesis is about *anything* and not
> about some one or more particular things, then our deductions
> constitute mathematics. Thus mathematics may be defined as
> the subject in which we never know what we are talking about,
> nor whether what we are saying is true.[4]

Mathematical existence parted company with physical existence. It
required only that the structure being invented on paper be free from
logical inconsistency. If this was the case then it was said to have
mathematical existence. Its properties could be studied by exploring
all the consequences of the prescribed rules. If a bad choice had been
made initially for the elements and rules of transformation of a mathematical structure so that they turned out to be inconsistent with each
other, then the structure was said not to exist mathematically.[5] Mathematical existence does not require that there be any part of physical
reality that follows the same rules, but if we believe Nature to be

rational then no part of physical reality could be described by a mathematically non-existent structure.

Incompleteness

One of the great challenges to mathematicians at the start of the twentieth century was to prove that mathematics was consistent and complete. This is, if one specified the axioms and rules of deduction in a mathematical system, then only true statements could be deduced in that system and the truth of any statement in the language of the system could have its truth or falsity decided by employing the rules of the system. It was confidently expected that this would be proved to be true. After all Hilbert had shown it to be true for Euclidean geometry and it was regarded as nothing more than a technical generalization to establish it for arithmetic as well. However, to general astonishment, Gödel showed that it wasn't true. There had to exist statements of arithmetic whose truth or falsity could not be decided. Arithmetic is incomplete.

It is possible to understand in a simple way how the argument works. Gödel established a correspondence between statements of mathematics and statements about mathematics (metamathematics). He did this by using prime numbers to encode each ingredient of a logical or mathematical statement. The product obtained by multiplying the prime numbers together then defines the whole statement. This number is now called its Gödel number. Moreover, since any number can be expressed as a product of prime numbers in one and only one way (for example, $51 = 3 \times 17$, $54 = 2 \times 3^3$, $9000 = 2^3 \times 3^2 \times 5^3$) the correspondence is unique: to each Gödel number there corresponds a logical statement. In this way every Gödel number corresponds to some logical statement about numbers (not necessarily a very interesting one) and each statement about numbers corresponds to some Gödel number.[6] For example, the Gödel number

$243\,000\,000 = 2^6 \times 3^5 \times 5^6$. The logical sentence is defined by the powers of the prime numbers taken in order, that is 656. The symbol 6 corresponds to the arithmetic symbol zero, 0, while 5 corresponds to equals, $=$; and so the Gödel number 656 represents the rather uninteresting arithmetical formula $0 = 0$.

Gödel's decisive step was to consider the statement

The theorem possessing Gödel number X is undecidable

He calculated its Gödel number and substituted that value for X in the statement. The result is a theorem that establishes its own unprovability.[7]

The essential features that make the incompleteness argument work is the possibility of self-reference – the correspondence between arithmetic and statements about arithmetic. This is only possible in logical systems which are complicated enough to allow statements about them to be coded uniquely and completely within the systems themselves; so that, if each possible ingredient of a logical statement is ascribed to a different prime number, then any complete statement can be represented by a Gödel number to which it corresponds. Some logical theories, like geometries, do not contain enough machinery to allow statements about themselves to be encoded within them in this way. These theories cannot display incompleteness.

Does it matter anyway?

One of the interesting things about this famous deduction is to ask whether it places real limits on scientific understanding of the universe. Gödel's monumental demonstration, that systems of mathematics have limits, gradually infiltrated the way in which philosophers and scientists viewed the world and our quest to understand it. Superficially, it appears that all human investigations of the universe must be limited. Science is based on mathematics; mathematics cannot discover all truths; therefore science cannot discover all truths. One of Gödel's contemporaries, Hermann Weyl, described Gödel's discovery as exercising 'a constant drain on the enthusiasm' with which he pursued his scientific research. He believed that this underlying pessimism, so different from the rallying cry with which Hilbert had issued to mathematicians in 1900, was shared 'by other mathematicians who are not indifferent to what their scientific endeavours mean in the context of man's whole caring and knowing, suffering and creative existence in the world'. In more recent times, a frequent writer on theology and science, Stanley Jaki, believes that Gödel prevents us from gaining an understanding of the cosmos as a necessary truth,

> Clearly then no scientific cosmology, which of necessity must be highly mathematical, can have its proof of consistency with itself as far as mathematics goes. In the absence of such consistency, all mathematical models, all theories of elementary particles, including the theory of quarks and gluons ... fall inherently short of being that theory which shows in virtue of it's a priori truth that the world can only be what it is and nothing else. This is true even if the theory happened to account with perfect accuracy for all phenomena of the physical world known at a particular time.[8]

and a fundamental barrier to understanding of the universe:

> It seems that on the strength of Gödel's theorem that the ultimate foundations of the bold symbolic constructions of mathematical physics will remain embedded forever in that deeper level of thinking characterized both by the wisdom and by the haziness of analogies and intuitions. For the speculative physicist this implies that there are limits to the precision of certainty, that even in the pure thinking of theoretical physics there is a boundary ... An integral part of this boundary is the scientist himself, as a thinker.[9]

In the last decade, Gödel's insights have been further illuminated by casting them into the language of information and randomness in the manner pioneered by Greg Chaitin.[10] This has created a different way of viewing the implications for physics. Science is the search for compressions of strings of data into briefer encodings ('laws of Nature') which contain the same information. We introduced this idea in our discussion of the power of simple mathematics. We said that any string of symbols which can be replaced by a formula or a rule that is shorter than the string itself is called *compressible*. Any string that cannot be abbreviated in this way is called *incompressible*. We can always demonstrate that a given string of symbols can be proved to be incompressible. The pattern needed to abbreviate the string of symbols might be one of those truths which cannot be proved. Thus, you can never know that your ultimate theory is the ultimate theory or not. There might always exist some deeper version of it: it might just be part of a larger and deeper theory.

These links between undecidability and randomness also allow us to forge further unexpected connections between Gödel and the efficiency of machines.[11] Undecidability will place limits on the efficiency of the machines of the far future. Suppose we take the example of a modern gas cooker. It is full of microprocessors, designed to sense the temperature inside the oven and implement instructions programmed into the control panel. The microprocessors store information temporarily until it is overwritten by new instructions. The more efficiently this new information can be encoded and stored in the microprocessor, the more efficiently the cooker operates, because it minimizes the unneeded work carried out erasing and overwriting the instructions lodged in its memory. But Chaitin's investigations show that Gödel's theorem is equivalent to the statement that we can never tell whether a program is the shortest one that will accomplish a given task. Hence, we can never find the most succinct program required to store the instructions for the operation of the cooker. As a result, the microprocessors we use will always overwrite more information than they need to: they will always possess some redundancy or ineffi-ciency. In practice, this 'logical friction' produces a decrease in gas cooker efficiency that is currently billions of times less than could be offset by simply cleaning it. Nonetheless, one day, these considera-tions might prove important to the operation of delicate nanotechno-logical machines and will be essential if we are to determine the ultimate capabilities of any technology.

Some human reactions

Intriguingly, and just to show the important role human psychology plays in assessing the significance of limits, some scientists, like Free-man Dyson, acknowledge that Gödel places limits on our ability to discover the truths of mathematics and science, but interpret this as ensuring that science will go on forever. Dyson sees the incomplete-ness theorem as an insurance policy against the scientific enterprise, which he admires so much, coming to a self-satisfied end; for

> Gödel proved that the world of pure mathematics is inex-haustible; no finite set of axioms and rules of inference can ever encompass the whole of mathematics; given any set of axioms, we can find meaningful mathematical questions which the axioms leave unanswered. I hope that an analogous situation exists in the physical world. If my view of the future is correct, it means that the world of physics and astronomy is also inex-haustible; no matter how far we go into the future, there will always be new things happening, new information coming in, new worlds to explore, a constantly expanding domain of life, consciousness, and memory.

Thus, we see the optimistic and the pessimistic responses to Gödel. The optimists, like Dyson, see his result as a guarantor of the never-ending character of human investigation. They see scientific research as part of an essential part of the human spirit which, if it were completed, would have a disastrous effect upon us. Karl Popper had this in mind when he wrote that 'continued growth is essential to the rational and empirical character of scientific knowledge; that if science ceases to grow it must lose that character.' The pessimists, like Jaki, by contrast, interpret Gödel as establishing that the human mind cannot know all (maybe not even most) of the secrets of Nature. They place more emphasis upon the possession and application of knowledge than on the process of acquiring it. The pessimist does not see the principal human benefit of science as arising from the quest for knowledge itself.

The same state of affairs elicits such diametrically opposed responses. On reflection we should not be too surprised. Many things in life create the same hiatus. It all depends whether you think your glass is half empty or half full. Gödel's own view was as unexpected as ever. He thought that intuition, by which we can 'see' truths of mathematics and science, was a tool that would one day be valued just as formally and reverently as logic itself:

> I don't see any reason why we should have less confidence in this kind of perception, i.e. in mathematical intuition, than in sense perception, which induces us to build up physical theories and to expect that future sense perceptions will agree with them and, moreover, to believe that a question not decidable now has meaning and may be decided in the future.[12]

Gödel was not minded to draw any strong conclusions for physics from his incompleteness theorems. He made no connections with the uncertainty principle of quantum mechanics, which was another great deduction which limited our ability to know, and which was discovered by Heisenberg just a few years before Gödel made his discovery. In fact, Gödel was rather hostile to any consideration of quantum mechanics at all. Those who worked at the same institute (no one really worked *with* him) believed that this was a result of his frequent discussions with Einstein who, in the words of John Wheeler (who knew them both) 'brainwashed Gödel' into disbelieving quantum mechanics and the uncertainty principle. Greg Chaitin records this account of Wheeler's attempt to draw Gödel out on the question of whether there is a connection between Gödel incompleteness and Heisenberg uncertainty.

> Well, one day I was at the Institute for Advanced Study, and I went to Gödel's office, and there was Gödel. It was winter and Gödel had an electric heater and had his legs wrapped in a blanket. I said 'Professor Gödel, what connection do you see between your incompleteness theorem and Heisenberg's uncertainty principle?' And Gödel got angry and threw me out of his office![13]

The claim that mathematics contains unprovable statements, physics is based on mathematics, therefore physics will not be able to discover everything that is true, has been around for a long time. More sophisticated versions of it have been constructed which exploit the possibility of incomputable mathematical operations being required to make predictions about observable quantities. From this vantage point, the mathematical physicist, Steven Wolfram, has conjectured that

> One may speculate that undecidability is common in all but the most trivial physical theories. Even simply formulated problems in theoretical physics may be found to be provably insoluble.[14]

Indeed, it is known that undecidability is the rule rather than the exception amongst the truths of arithmetic.[15]

Checking the fine print

With these worries in mind, let us look a little more closely at what Gödel's result might have to say about the course of physics. The situation is not so clear-cut as the commentators would have us believe. It is useful to lay out the precise assumptions that underlie Gödel's deduction of incompleteness. Gödel's theorem says that if a formal system is

1. *finitely specified*
2. *large enough to include arithmetic*
3. *consistent*

then it is *incomplete*.

Condition 1 means that there is not an incomputable infinity of axioms. We could not, for instance, choose our system to consist of all the true statements about arithmetic because this collection cannot be finitely listed in the required sense. Condition 2 means that the formal system includes all the symbols and axioms used in arithmetic. These symbols are 0, 'zero', S, 'successor of', +, ×, and =. Hence, the number two is the successor of the successor of zero, written as the *term* SS0, and two and plus two equals four is expressed as SS0 + SS0 = SSSS0.

The structure of arithmetic plays a central role in the proof of SS0's own theorem. Special properties of numbers, like their primeness and the fact that any number can be expressed in only one way as the product of the prime numbers that divide it, were used by Gödel to establish the vital correspondence between statements of mathematics and statements about mathematics. Thereby, linguistic paradoxes like that of the 'liar' could be embedded, like Trojan horses, within the structure of mathematics itself. Only logical systems which are rich enough to include arithmetic allow this incestuous encodings of statements about themselves to be made within their own language.

Again it is instructive to see how these requirements might fail to be met. If we picked a theory that consisted of references to (and relations between) only the first ten numbers (0,1,2,3,4,5,6,7,8,9) then Condition 2 fails and such a mini-arithmetic is complete. Arithmetic makes statements about individual numbers, or terms (like SS0 above). If a system does not have individual terms like this but, like Euclidean geometry, only makes statements about points, circles, and lines, in general, then it cannot satisfy Condition 2. Accordingly, as Alfred Tarski first showed, Euclidean geometry is complete. There is nothing magical about the flat, Euclidean nature of the geometry either: the non-Euclidean geometries on curved surfaces are also complete. Similarly, if we had a logical theory dealing with numbers that only used the concept of 'greater than' without referring to any specific numbers then it would be complete: we can determine the truth or falsity of any statement about real numbers involving the 'greater than' relationship.

Another example of a system that is smaller than arithmetic is arithmetic without the multiplication, ×, operation. This is called Presburger arithmetic (the full arithmetic is called Peano arithmetic after the mathematician who first expressed it axiomatically, in 1889). At first this sounds strange; in our everyday encounters with multiplication it is nothing more than a shorthand way of doing addition (2 + 2 + 2 + 2 + 2 + 2 = 2 × 6, but in the full logical system of arithmetic, in the presence of logical quantifiers like 'there exists' or 'for any', multiplication permits constructions which are not merely equivalent to a succession of additions.

Gödel showed, as part of his doctoral thesis work, that Presburger arithmetic is complete: all statements about the addition of natural numbers can be proved or disproved; all truths can be reached from the axioms.[16] Similarly, if we create another truncated version of arithmetic, which does not have addition, but retains multiplication, this is also complete. It is only when addition and multiplication are simultaneously present that incompleteness emerges. Extending the system further by adding extra operations like exponentiation to the repertoire of basic operations makes no difference. Incompleteness remains but no intrinsically new form of it is found. Arithmetic is the watershed in complexity.

The use of Gödel to place limits on what a mathematical theory of physics (or anything else) can ultimately tell us seems a fairly straight-forward consequence. But as one looks more carefully into the ques-tion, things are not quite so simple. Suppose, for the moment, that all the conditions required for Gödel's theorem to hold are in place. What would incompleteness look like in practice? We are familiar with the situation of having a physical theory which makes accurate pre-dictions about a wide range of observed phenomena: we might call it 'the standard model'. One day, we may be surprised by an observation about which it has nothing to say. It cannot be accommodated within its framework. Examples are provided by some so-called 'grand uni-fied theories' in particle physics. Some early editions of these theories had the property that all neutrinos must have zero mass. Now if a neutrino is observed to have a non-zero mass (as everyone believes it will have, and some experiments have even claimed to have measured) then we know that the new situation cannot be accommodated within our original theory. What do we do? We have encountered a certain sort of incompleteness, but we respond to it by extending or modifying the theory to include the new possibilities. Thus, in practice, incom-pleteness looks very much like inadequacy in a theory.

In the case of arithmetic, if some statement about arithmetic is known to be undecidable (there are known statements of this sort; it means that both their truth and falsity are consistent with the axioms of arithmetic) then we have two ways of extending the structure. We can create two new arithmetics: one which adds the undecidable statement as an extra axiom, the other which adds its negation as a new axiom. Of course, the new arithmetics will still be incomplete, but they can always be extended to accommodate any incomplete-ness. Thus, in practice, a physical theory can always be enlarged by adding new principles which force all the undecidability into the part of the mathematical realm which has no physical manifestation.

Incompleteness would then always be very hard, if not impossible, to distinguish from incorrectness or inadequacy.

Escape clauses

An interesting example of this dilemma is provided by the history of mathematics. During the sixteenth century, mathematicians started to explore what happened when they added together infinite lists of numbers. If the quantities in the list get larger then the sum will 'diverge'; that is, as the number of terms approaches infinity so does the sum. An example of this is the sum

$$1 + 2 + 3 + 4 + 5 + \ldots = \text{infinity.}$$

However, if the individual terms get smaller and smaller suffi-ciently rapidly,[17] then the sum of an infinite number of terms can get closer and closer to a finite limiting value which we shall call the sum of the series; for example

$$1 + \tfrac{1}{4} + \tfrac{1}{9} + \tfrac{1}{16} + \tfrac{1}{25} + \tfrac{1}{36} + \tfrac{1}{49} + \ldots = \frac{\pi^2}{6} = 1.644934067\ldots$$

This left mathematicians to worry about a most peculiar type of unending sum,

$$1 - 1 + 1 + - 1 + 1 + - 1 + 1 - \ldots = ?????$$

If you divide up the series into pairs of terms it looks like $(1-1) + (1-1) + \ldots$ and so on. This is just $0 + 0 + 0 + \ldots =)$ and the sum is zero. But think of the series as $1 - \{1 - 1 + 1 - 1 + 1 - \ldots\}$ and it looks like $1 - \{0\} = 1$. We seem to have proved that $0 = 1$.

Mathematicians had a variety of choices when faced with ambigu-ous sums like this. They could reject infinities in mathematics and deal only with finite sums of numbers. Or, as Cauchy showed in the early nineteenth century, the sum of a series like the last one must be defined by specifying more closely what is meant by its sum. The limiting value of the sum must be specified together with the procedure used to calculate it. The contradiction $0 = 1$ arises only when one omits to specify the procedure used to work out the sum. In both cases it is different and so the two answers are not the same. Thus, here we see a simple example of how a limit is side-stepped by enlarging the concept which seems to create limitations. Divergent series can be dealt with consistently so long as the concept of a sum for a series is suitably extended.[18]

Another possibility is that the physical world only makes use of the decidable part of mathematics. We know that mathematics is an infinite sea of possible structures. Only some of those structures and patterns appear to find existence and application in the physical world. It may be that they are all from the subset of decidable truths. Things may be better protected even than that: perhaps only computable patterns are instantiated in physical reality?

It is also possible that the conditions required to prove Gödel's incompleteness do not apply to physical theories. Condition 1 requires the axioms of the theory to be listable. It might be that the laws of physics are not listable in this predictable sense. This would be a radical departure from the situation that we think exists, where the number of fundamental laws is believed to be not just listable, but finite (and very small). But it is always possible that we are just scratching the surface of a bottomless tower of laws, only the top of which has significant effects upon our experience. However, if there were an unlistable infinity of physical laws then we would face a more formidable problem than that of incompleteness.

In fact, in 1940, Gerhard Gentzen, one of Hilbert's young students who lost his life in the war soon afterwards, showed that it was possible to circumvent Gödel's conclusions and deduce all the truths of arithmetic if a procedure of transfinite induction is included.

Again, the operations of Nature might include such a non-finite system of axioms. We are inclined to think of incompleteness as something undesirable because it implies that we will not be able to 'do' something. But we could turn the situation on its head and conclude that Nature is consistent and complete but cannot be captured by a finite set of axioms. There is something aesthetically satisfying about this superhuman complexion to things.

An equally interesting issue is that of finiteness. It may be that the universe of physical possibilities is finite, although astronomically large. However, no matter how large the number of primitive quantities to which the laws refer, so long as they are finite the resulting system of inter-relationships will be complete. We should stress that although we habitually assume that there is a continuum of points of space and time this is just an assumption that is very convenient for the use of simple mathematics. There is no deep reason to believe that space and time are continuous, rather than discrete, at their most fundamental microscopic level; in fact, there are some theories of quantum gravity that assume they are not. Quantum theory has introduced discreteness and finiteness in a number of places where once we believed in a continuum of possibilities. Curiously, if we give up this continuity, so that there is not necessarily another point in between any two sufficiently close points you care to choose, space-time structure becomes vastly more complicated. Many more complicated things can happen. This question of finiteness might also be bound up with the question of whether the universe is finite in volume and whether the number of elementary particles (or whatever the most elementary entities might be) of Nature are finite or infinite in number. Thus there might only exist a finite number of terms to which the ultimate logical theory of the physical world applies. Hence, it would be complete.

An interesting possibility with regard to the application of Gödel to the laws of physics is that Condition 2 of the incompleteness theorem might not be met. How could this be? Although we seem to make wise use of arithmetic, and much larger mathematical structures, when we carry out scientific investigations of the laws of Nature, this does not mean that the inner logic of the physical universe needs to employ such a large structure. It is undoubtedly convenient for us to use large mathematical structures together with concepts like infinity, but this may be an anthropomorphism. The deep structure of the universe may be rooted in a much simpler logic than that of full arithmetic, and hence be complete. All this would require would be for the underlying structure to contain either addition or multiplication but not both. Recall that all the sums that you have ever done have used multiplication simply as a shorthand for addition. They would be possible in Presburger arithmetic as well. Alternatively, a basic structure of reality that made use of simple relationships of a geometrical variety, or which derived from 'greater than' or 'less than' relationships, or subtle combinations of them all could also remain complete.[19] The fact that Einstein's theory of general relativity replaces many physical notions like force and weight by *geometrical* distortions in the fabric of space-time may well hold some clue about what is possible here.

The laws of physics might be fully expressible in terms of a mathematical system that is complete, but in practice we would always be far more concerned with making sure that we had got the *correct* system than a complete system.

There is another important aspect of the situation to be kept in view. Even if a logical system is complete, it always contains unprovable 'truths'. These are the axioms which are chosen to define the system. And after they are chosen, all the logical system can do is deduce conclusions from them. In simple logical systems, like Peano arithmetic, the axioms seem reasonably obvious because we are thinking backwards – formalizing something that we have been doing intuitively for thousands of years. When we look at a subject like physics, there are parallels and differences. The axioms, or laws, of physics are the prime target of physics research. They are by no means intuitively obvious, because they govern regimes that can lie far outside of our experience. The outcomes of those laws are unpredictable in certain circumstances because they involve symmetry breakings. Trying to deduce the laws from the outcomes is not something that we can ever do uniquely and completely by means of a computer program.

Thus, we detect a completely different emphasis in the study of formal systems and in physical science. In mathematics and logic, we start by defining a system of axioms and laws of deduction. Then, we might try to show that the system is complete or incomplete, and deduce as many theorems as we can from the axioms. In science, we are not at liberty to pick any logical system of laws that we choose. We are trying to find the system of laws and axioms (assuming there is one – or more than one, perhaps) that will give rise to the outcomes that we see. As we stressed earlier, it is always possible to find a system of laws which will give rise to any set of observed outcomes. But it is the very set of unprovable statements that the logicians and the mathematicians ignore – the axioms and laws of deduction – that the scientist is most interested in discovering rather than simply assuming. The only hope of proceeding as the logicians do, would be if for some reason there is only one possible set of axioms or laws of physics. So far, this does not seem likely;[20] even if it were we would not be able to prove it.

Concrete examples

Specific examples have been given of physical problems which are undecidable. As one might expect from what has just been said, they do not involve an inability to determine something fundamental about the nature of the laws of physics or the most elementary particles of matter. Rather, they involve an inability to perform some specific mathematical calculation, which inhibits our ability to determine the course of events in a well-defined physical problem.

However, although the problem may be mathematically well-defined, this does not mean that it is possible to create the precise conditions required for the undecidability to exist.

An interesting series of examples of this sort has been created by the Brazilian mathematicians Francisco Doria and N. da Costa.[21] Responding to a challenge problem, posed by the Russian mathematician Vladimir Arnold, they investigated whether it was possible to have a general mathematical criterion which would decide whether or not any equilibrium was stable. A stable equilibrium is a situation like a ball sitting in the bottom of a basin – displace it slightly and it returns to the bottom; an unstable equilibrium is like a needle balanced vertically – displace it slightly and it moves away from the vertical.[22] When the equilibrium is of a simple nature this problem is very elementary; first-year science students learn about it. But, when the equilibrium exists in the face of more complicated couplings between the different competing influences, the problem soon becomes more complicated than the situation studied by science students. So long as there are only a few competing influences the stability of the equilibrium can still be decided by inspecting the equations that govern the situation. Arnold's challenge was to discover an algorithm which tells us if this can always be done, no matter how many competing influences there are, and no matter how complex their inter-relationships. By 'discover' he meant find a formula into which you can feed the equations which govern the equilibrium along with your definition of stability, and out of which will pop the answer 'stable' or 'unstable'.

Strikingly, da Costa and Doria discovered that there can exist no such algorithm. There exist equilibria characterized by special solutions of mathematical equations whose stability is undecidable. In order for this undecidability to have an impact on problems of real interest in mathematical physics the equilibria have to involve the interplay of very large numbers of different forces. While such equilibria cannot be ruled out, they have not arisen yet in real physical problems. Da Costa and Doria went on to identify similar problems where the answer to a simple question, like 'Will the orbit of a particle become chaotic?', is Gödel undecidable. Others have also tried to identify formally undecidable problems. Geroch and Hartle have discussed problems in quantum gravity that predict the values of potentially unobservable quantities as a sum of terms whose listing is known to be a Turing uncomputable operation.[23] Pour-El and Richards[24] showed that very simple differential equations, which are widely used in physics, like the wave equation, can have uncomputable outcomes when the initial data is not very smooth. This lack of smoothness gives rise to what mathematicians call an 'ill-posed' problem. It is this feature that gives rise to the uncomputability. However, Traub and Wozniakowski have shown that every ill-posed problem is well-posed on the average under rather general conditions.[25] Wolfram[26] gives examples of intractability and undecidability arising in condensed matter physics.

The study of Einstein's general theory of relativity also produces an undecidable problem if the mathematical quantities involved are unrestricted.[27] When one finds an exact solution of Einstein's equations it is always necessary to discover whether it is just another, known solution that is written in a different form. Usually, one can investigate this by hand, but for complicated solutions computers can help. For this purpose we require computers unprogramed to algebraic manipulations. Such a computer can check various quantities to discover if a given solution is equivalent to one already sitting in its memory bank of known solutions. In practical cases encountered so far, this checking procedure comes up with a definite result after a small number of steps. But in general the comparison is an undecidable process equivalent to another famous undecidable problem of pure mathematics, 'the word problem' of group theory.

Finally ...

We have reviewed some of the aspects of mathematical explanation that are novel: its ubiquity in explaining the world around us, the unbounded character of mathematical existence (which requires nothing more than logical self-consistency), and the disconcerting incompleteness of logical structures that are rich enough to contain arithmetic. We turned then to look at some of the speculations that have been made about the limits that might exist on a mathematical understanding of the universe because of this incompleteness. We discovered that there is a critical divide in complexity in mathematical structures that signals the appearance of incompleteness. That critical level is linked to the possibility of establishing a form of self-reference. Arithmetic is the simplest structure that allows that form of individuality to exist.[28]

Acknowledgements

I would like to thank John Cornwell and Jesus College for their invitation to participate in the stimulating meeting, and Peter Lipton, Martin Rees, Doug Robertson, Bill Saslaw, and John Cornwell for discussions.

Notes

1. J. D. Barrow. *Pi in the sky: counting, thinking and being* (Oxford University Press, 1992).

2. The philosopher Immanuel Kant argued that Euclidean geometry was the only geometry that is humanly thinkable. It was forced upon us like a strait-jacket by the way minds work. This was soon shown to be totally incorrect by the creation of new geometries. In fact, Kant should not have needed new mathematical developments to tell him this. By looking at any Euclidean geometrical example (for example a triangle

on a flat surface) in a curved mirror it should have been clear that the laws of reflection guarantee that there must exist geometrical 'laws' on the curved surface which are reflections of those that exist on the flat surface.

3. J. Richards, 'The reception of a mathematical theory: non-Euclidean geometry in England 1868–1883', in *Natural order: historical studies of scientific culture* (ed. B. Barnes and S. Shapin), pp. 143–66 (Sage, Beverly Hills, 1979); J. L. Richards, *Mathematical visions: the pursuit of geometry in Victorian England* (Academic, Boston, 1988); E. A. Purcell, *The crisis of democratic theory* (University of Kentucky Press, Lexington, 1973); Barrow, *Pi in the Sky*, op cit.

4. B. Russell. 'Recent work on the principles of mathematics'. *International Monthly* (1901), 4.

5. Unexpectedly, the Austrian mathematician Kurt Gödel showed that if a mathematical structure is rich enough to contain arithmetic then it is not possible to prove that its defining axioms are inconsistent. If they are assumed to be consistent then the structure is necessarily incomplete in the sense that there must exist statements framed in the language of the structure which can neither be proved to be true nor false using the rules of reasoning of the system. Euclidean and non-Euclidean geometries are not rich enough to contain the structure of arithmetic and so the incompleteness theorem does not apply to them. For more details, see Chapter 8 in J. D. Barrow, *Impossibility: the limits of science and the science of limits* (Oxford University Press, 1998).

6. The requirement that statements make logical sense means that not every integer is a Gödel number.

7. At the start of this chapter we looked at paradoxical statements which asserted their own falsity ('This statement is false'). The logician Leon Henkin has given his name to statements which are automatically true ('This sentence is provable'). Henkin sentences are self-proving sentences. The proof was given by Löb, who proved the following interesting theorem relating statements and metastatements: If we have a system, S, in which the statement '*if A is provable in S then A is true*' is provable in S, then A is provable in S. This theorem of Löb's implies one of Gödel's incompleteness theorems as a particular example, if we take the formula $0 = 1$ to be the statement labelled A. Then, we can conclude that the consistency of S is not provable in S. For further discussion about incompleteness theorems, see C. Smorynski, 'The incompleteness theorems', in *Handbook of mathematical logic* (ed. J. Barwise), pp. 821–65 (North Holland, Amsterdam, 1977). For a serious popularization, see R. Smullyan, *Forever undecided: a puzzle guide to Gödel* (Oxford University Press, 1987).

8. S. Jaki. *Cosmos and creator* (Scottish Academic Press, Edinburgh, 1989), p. 49.

9. S. Jaki. *The relevance of physics* (University of Chicago Press, 1966), p. 129.

10. G. Chaitin. *Information, randomness and incompleteness* (World Scientific, Singapore, 1987).

11. S. Lloyd. 'The calculus of intricacy'. *The Sciences* (Sept/Oct 1990), 38–44; Barrow, *Pi in the Sky*, pp. 130–40.

12. K. Gödel. 'What is Cantor's Continuum Problem?' In *Philosophy of mathematics* (ed. P. Benacerraf and H. Putnam), p. 43 (Cambridge University Press, 1983).

13. Recorded by Chaitin. See J. Bernstein, *Quantum Profiles* (Princeton University Press, 1991), pp. 140–1, and K. Svozil, *Randomness and undecidability in physics* (World Scientific, Singapore, 1993) p. 112.

14. S. Wolfram. *Cellular automata and complexity: collected papers* (Addison Wesley, Reading, MA, 1994).

15. C. Calude, *Information and randomness: an algorithmic perspective* (Springer, Berlin, 1994); C. Calude, H. Jürgensen, and M. Zimand, 'Is independence an exception?', *Appl. Math. Comput.* (1994), **66**, 63; K. Svozil, in *Boundaries and barriers: on the limits of scientific knowledge* (ed. J. Casti and A. Karlqvist), p. 215 (Addison Wesley, New York, 1996).

16. The decision procedure is in general double-exponentially long, though. That is, the computational time required to carry out N operations grows as $(2^N)^N$. Presburger arithmetic allows us to talk about positive integers, and variables whose values are positive integers. If we enlarge it by permitting the concept of sets of integers to be used, then the situation becomes almost unimaginably intractable. It has been shown that this system does not admit even a K-fold exponential algorithm, for any finite K. The decision problem is said to be non-elementary in such situations. The intractability is unlimited.

17. That the terms in the sum get progressively smaller is a necessary but not a sufficient condition for an infinite sum to be finite. For example, the sum $1 + \frac{1}{2} + \frac{1}{3} + \frac{1}{4} + \frac{1}{5}$ ... is infinite.

18. R. Rosen. 'On the limitations of scientific knowledge'. In Casti and Karlqvist, *Boundaries and barriers*, p. 199.

19. John A. Wheeler has speculated about the ultimate structure of space-time being a form of 'pregeometry' obeying a calculus of propositions restricted by Gödel incompleteness. We are proposing that this pregeometry might be simple enough to be complete; see C. Misner, K. Thorne, and J. A. Wheeler, *Gravitation*, pp. 1211–12 (W. H. Freeman, San Francisco, 1973).

20. The situation in superstring theory is still very fluid. There appear to exist many different, logically self-consistent superstring theories, but there are strong indications that they may be different representations of a much smaller number (maybe even just one) theory.

21. N. C. da Costa and F. Doria, *Int. J. Theor. Phys.* (1991), **30**, 1041; *Found. Phys. Letts.* (1991), **4**, 363.

22. Actually, there are other more complicated possibilities clustered around the dividing line between these two simple possibilities and it is these that provide the indeterminacy of the problem in general.

23. R. Geroch and J. Hartle. 'Computability and physical theories'. *Foundations of Physics* (1986), **16**, 433. The problem is that the calculation of a wave function for a cosmological quantity involves the sum of quantities evaluated on every four-dimensional compact manifold in turn. The listing of this collection of manifolds is uncomputable.

24. M. B. Pour-El and I. Richards, 'A computable ordinary differential equation which possesses no computable solution', *Ann. Math. Logic* (1979), **17**, 61; 'The wave equation with computable initial data such that its unique solution is not computable', *Adv. Math.* (1981), **39**, 215; 'Non-computability in models of physical phenomena', *Int. J. Theor. Phys.* (1982), **21**, 553.

25. J. F. Traub and H. Wozniakowski. 'Information-based complexity: new questions for mathematicians'. *Mathematical Intelligencer* (1991), **13**, 34.

26. S. Wolfram, 'Undecidability and intractability in theoretical physics', *Phys. Rev. Lett.* (1985), **55**, 449; 'Physics and computation', *Int. J. Theo. Phys.* (1982), **21**, 165.

27. If the metric functions are polynomials then the problem is decidable, but is computationally double-exponential. If the metric functions are allowed to be sufficiently smooth then the problem becomes undecidable; see the article on 'Algebraic simplification' by B. Buchberger and R. Loos, in B. Buchberger, R. Loos, and G. Collins, *Computer algebra: symbolic and algebraic computation*, pp. 11–44, 2nd edn (Springer, Vienna, 1983). I am grateful to Malcolm MacCallum for supplying these details.

28. For a fuller discussion of the whole range of limits that may exist on our ability to understand the universe, see the author's discussion in J. D. Barrow, *Impossibility* (Oxford University Press, 1998), on which the discussion here is based.

Peter Atkins

6

Ponderable matter: explanation in chemistry

THE most extraordinary event in the evolution of chemical explanation occurred in the eighteenth century. Until then matter, though tangible, had been inexplicable. Although matter was everywhere, both inside us and outside us, and though matter could be transformed in a variety of natural and contrived ways, it was unrationalized; in the physical sense it was ponderous but in a mental sense imponderable. Though copper might occur with zinc, and though the manipulation of blends of the two metals provided the launching pad for civilization when bronze displaced stone for utensil, decoration, and weapon, there was no inkling of why they were truly cousins. Though grapes could be fermented and the fermented product turn to vinegar, there was observation without comprehension. Despite their almost infinite meddlings, the alchemists never understood their failure, except by self-deception and simple deception, to transmute base metal into gold.

The key that turned the lock of understanding, the conversion of the imponderable to the ponderable, was, appropriately, the balance. When Antoine Lavoisier brought the chemical balance to bear on matter, he turned a qualitative body of knowledge into a physical science. In the crucial step, alchemy effectively became chemistry, for from then on number could be attached to matter and the rigour of quantitative scientific investigation used to tease out understanding. Over the following two centuries, chemists assembled insight in to the nature of matter: Dalton justified atoms, Rutherford discerned something of their structure, rendered the understanding of atoms precise. Now we understand matter – at the level characteristic of chemistry, where relatively stable combinations of the fundamental particles still being explored by physics have persistent personalities.

Chemistry is concerned with the structure of matter (at the level referred to above) and the changes that matter can undergo. The two central planks of its explanations are atoms and energy, and we shall regard these two concepts as our currency of discourse. As we shall see, although modern chemistry does have some predictive power, its greater strength lies in its power to rationalize. That is the nature of a subject lying across the frontier where simplicity – atoms – meets complexity bulk matter. Indeed, as chemistry has marched into the molecular scale, in the form of biological macromolecules such as proteins and nucleic acids, once again it has had at least partially to cede prediction to rationalization.

Although the Greeks had speculated that matter is composed of atoms, they had speculated about most things and in most ways, and it is hardly surprising that one of their guesses had in due course turned out to be right. The first real evidence for the existence of an end to cutting up matter came from Dalton's considerations of the results of weighing matter before and after it had been transformed by a chemical reaction. He found that a particular mass of one substance combined with a particular mass of another, and inferred that matter was corpuscular with atoms of different elements having characteristic masses. Dalton's recognition that a chemical reaction is a process in which atoms merely change partners, with reactants one combination of atoms and products another combination of the same atoms, is still the all-pervasive foundation of explanation in chemistry.

Dalton's hypothesis of the existence of atoms and his ascription to them of different masses was an inference from many experiments based on weighing. Inference of an entirely different order is now open to use in the form of scanning tunnelling microscopy, which is a technique that, after a certain amount of manipulation of the data, yields images that strikingly confirm the existence of atoms. These atoms appear as bumps lying in serried ranks to form surfaces, and molecules lying on such surfaces can also be discerned as further groups of bumps making patterns in accord with the known architecture of molecules. In so far as we can call this kind of visual portrayal of surfaces 'seeing', we can now state that we know that atoms exist because we can see them.

The existence of atoms with characteristic masses is the concept that underlies that other great icon of chemistry, the periodic table. The periodic table is not so much an explanation of chemistry as a helpful organization of the elements to show their relationships to one another. First, the existence of atoms clarifies what chemists mean by an 'element'. Until the existence of atoms was established, an element had to be defined as a substance that thwarted a chemist's attempts to break it down into something simpler. Chemists increased the aggression of their onslaughts, using first heat then electricity. Nuclear physicists, were they allowed into the fold, would slaughter even more effectively, so the 'elementary' character of certain substances – all substances – would gradually change. By defining an element as a substance composed of only one kind of atom, the definition was sharpened and no longer twisted in the wind of progress.

Elements, of course, are not elementary. They are composed of other entities – as it became clear after 1897 when the electron was discovered, and as it should have been clear all along, because the very complex properties of atoms of the elements (for instance, their abilities to enter only into specific types of combination) cannot stem from entities without structure, and the existence of structure implies the present of smaller components. However, elements are sufficiently elementary for a large part of chemistry: they are bundles of subatomic particles with sufficient stability to constitute recognizable matter in the everyday world, and are sufficiently stable – and this is the crucial point – to survive in most cases the ravages of chemical reactions. They are useful, persistent packages of subatomic particles.

The periodic table portrays the personalities of these persistent packages of particles. Patterns in matter had been sought for decades before Mendeleev's dream brought order to them. He was aided by two things. First, chemists had identified an abundance of elements. When only a handful of elements were known, the future periodic table was little more than an apparently random atlas of isolated islands, and no pattern was discernible. By the middle of the nineteenth century, chemists had refined their art and over three score elements were known. The islands of the future periodic table now started to have neighbours and to run together: it was as though water was running out of the table, revealing the land beneath and showing that islands were in fact continents. The second feature that aided Mendeleev was his knowledge of reasonably reliable atomic weights, so he could organize the elements by referring to a quantitative feature of the element rather than relying solely on intuition and chemical insight. So, as the story goes, Mendeleev fell asleep after working obsessively but unsuccessfully on the problem of organizing the elements for the textbook he was writing; on awaking, the table was in place.

Although the periodic table is an icon of chemistry and viewed proudly by chemists, we must be cautious about ascribing to it much explanatory power. It is a wonderful instrument for learning chemistry, for it brings order out of apparent chaos. But few chemists gaze at it for inspiration in their research, although it is generally useful for suggesting analogies and likely properties. For explanation of properties, we have to look for the physical reasons underlying the rhythms of the table.

For explanation of the periodicity of the elements chemists turn to the electronic structure of the atoms and of the elements. Herein lies their greatest debt to quantum mechanics, for the periodic table and hence chemistry would be inexplicable without quantum theory. The explanation of the periodic table is understood by chemists at two levels of description, one qualitative the other quantitative. The qualitative explanation of the periodic table is based on an extrapolation from the electronic structure of hydrogen, the only atom for which analytic solutions of the Schrödinger equation can be obtained. In their vision of the hydrogen atom, chemists think of the atom as possessing an infinite number of orbitals which its electron can occupy. These orbitals correspond to characteristic energies and angular momenta. In a hydrogen atom, the single electron occupies just one of these orbitals, the one of lowest energy and, as it happens, zero orbital angular momentum. Chemists take the same orbitals, suitably modified to account for the different nuclear charge of heavier atoms, as the basis for their description of all other atoms. In these atoms, the numerous electrons (92, for instance, in an atom of uranium) also occupy the available orbitals in such a way that the atom achieves the lowest energy subject to one of the most amazing features of matter: this restriction is the Pauli principle, one of the most profound of all principles governing matter and still not truly understood. (But that is characteristic of quantum mechanics as a whole: although no one really understands it, everyone can use it.)

A few simple ideas about the role of electron–electron repulsions and their effect on the relative energies of orbitals coupled with the Pauli principle is sufficient to explain the structure of the periodic table and hence the periodic relationships between the elements. These qualitative ideas are refined and generally supported by detailed quantitative calculations based on the numerical solution of the Schrödinger equation, and chemists can claim to have as good a quantitative understanding of the structures of atoms as is necessary for their trade. This confidence begins to fade a little for really heavy elements and for electrons close to nuclei, when relativistic effects need to be considered. Thus, it is not possible to account for the colour of gold or the liquid character of mercury without invoking relativistic effects, but even these much more difficult calculations are now fully understood and implemented.

However, here there lies a subtle worry. Although chemists express most of their structural explanations in terms of electrons in orbitals, the sophisticated calculations used to support their speculative insights are based on the view that orbitals don't really exist. The resolution of this paradox is the same as in other areas of science: the concept of an occupied orbital captures some essential feature of the distribution of electrons in atoms and their ability to lose and gain other electrons (which, as we shall now see, is at the root of bond formation).

The electronic structures of atoms is only the start of a chemist's rationalization of chemistry. Chemistry is all about the compounds that atoms form as they link together with different partners and form structures ranging from the trivial (hydrogen molecules) to the most potent and profound (DNA). The understanding of bonding that emerged and was effectively completed during the first half of the twentieth century is an excellent example of the growth and refinement of explanation in chemistry, and particularly of the models that are used as an aid to visualization, calculation, and exposition.

The origins of an explanation of bonding (if we forget the hooks and eyes that the mechanically minded Victorians supposed existed on atoms) emerged soon after it was realized that electrons occupied the extranuclear regions of atoms and that they could be dislodged with relatively little input of energy. Once the nuclear atom had been identified (by Rutherford), it was realized that the plastic, mutable region of atoms was the region occupied by the extranuclear electrons, and that chemistry is effectively the manifestation of the properties of that region. In a suggestion of extraordinary insight, well before quantum mechanics had been developed and electronic structure elucidated, G. N. Lewis proposed (in 1916) that a chemical bond was essentially an electron pair shared by the two bonded atoms. That is, the nuclei of bonded atoms are stuck together by a pair of electrons that lay between them. Lewis had utterly no idea of why two electrons should be so important, but with this suggestion he was able to rationalize a whole swathe of molecular structures and to relate the bonding characteristics of atoms to their location in the periodic table.

The interpretation of Lewis's proposal and an explanation of the importance of the electron pair had to await the development of quantum mechanics (in 1927) and its application to molecules a few years later. Two rival quantum theories of the chemical bond, valence-bond theory and molecular orbital theory, were proposed almost simultaneously. At first sight, the two approaches are quite different, for the former focuses on pairs of atoms in a molecule and builds up a description of bonding in terms of effects local to pairs of atoms, whereas the latter treats the molecule as a whole and allows the bonding effects of electrons to be dispersed over the entire molecule. Both approaches can be brought into line by progressive improvement, so there is no actual conflict in their explanation of bonding, and characteristics of a particular bond or who is focussing on a reaction in which a particular group of atoms is being displaced from a molecule by an incoming attacking atom will tend to favour the language of valence-bond theory whereas a chemist emphasizing a global property of a molecule, such as its optical properties, will favour molecular orbital theory. Broadly speaking, organic chemists use molecular orbital theory. In certain cases, the two cultures meet in the same molecule: the bonding in benzene, for instance, is commonly discussed in a mixture of both languages. The conclusion to draw is that chemists are pragmatic with their explanations, dressing in one suit of clothes when it suits them, and another when an alternative is more appropriate. That is versatility, not conflict, and as always, versatility can be the source of inspiration.

For technical reasons related to the ease of implementation of the appropriate calculations, molecular orbital theory has been vastly more highly developed than valence-bond theory, and except in isolated pockets of resistance is virtually the sole technique of computation. That should not be taken to imply that molecular orbital calculations are simple: computational chemists, meteorologists, and cryptanalysts are the heaviest users of supercomputers. Nevertheless an enormous effort goes into the calculation of molecular structures for they can be screened for, among other properties, pharmacological effectiveness. Thus it may be that although computational chemistry has not yet saved a human life, it has saved a thousand rats.

The development of a quantum mechanical explanation of the chemical bond brought in its train the refutation of certain of Lewis's ideas. However, as happened for atoms and as happens elsewhere in science when models are rendered more sophisticated, the *essence* of his ideas could be rationalized and could be regarded as surviving. In a sense, Lewis's model is a form of accounting that gives an outcome that is usually consistent with the results of molecular orbital and valence-bond theory. And it is still used, not only in introductory chemistry, where putting pairs of electrons between pairs of atoms to represent the bonds and applying certain simple rules about how many such pairs can be placed (a step that reflects the location of the element in the periodic table) is an excellent way of rationalizing the existence of known compounds and dismissing certain speculations. The more sophisticated use of Lewis's approach is in organic chemistry, where the migration of electron pairs around a molecule and the provision of a new pair by an incoming reagent is used to rationalize the course of reactions and in some cases to make successful prediction. It is highly regrettable that Lewis, who did so much to provide the foundations of explanation in twentieth-century chemistry, never received the Nobel Prize: never was one not awarded more deserved.

Lewis's name is also associated with the classification of reactions, particularly those between acids and bases. One of the paradigmatic reactions of chemistry is the reaction between an acid (such as hydrochloric acid) and a base (such as ammonia) to give a salt (ammonium chloride) and water. But what is an acid and what is a base?

The first plausible attempt at explanation and definition (if we discount speculations about particles with sharp spikes for acids) was proposed by Svante Arrhenius, who suggested that an acid was a compound that released hydrogen ions (H^+) in water and that a base was something that generated hydroxide ions (OH^-) in water. That was all very well, but the definition was seen to be too restrictive once chemists began to carry out reactions in solvents other than water, and in the absence of any solvent, and found similar patterns of reaction.

To enlarge the universe of acid–base chemistry, Johannes Brønsted and Thomas Lowry proposed independently that a species that could donate a hydrogen ion should be regarded as an acid and that a species that could accept a hydrogen ion should be regarded as a base. In this way, the whole of Arrhenius chemistry survived, but the pattern of that chemistry now embraced a much wider range of species and environments. Note how the explanation of acid and base character has moved away from the reactions of compounds in water to the ability of species to donate and accept a hydrogen ion.

At about the same time, Lewis proposed an even wider definition, in which an acid was defined as a species that could accept an electron pair and a base was defined as a species that could donate an electron pair. This elaboration of the definition expanded the pattern of acid–base chemistry even further, for the reaction no longer had to involve hydrogen ions. Nevertheless, all the Brønsted–Lowry chemistry survived (and therefore the Arrhenius chemistry did too), for a hydrogen ion – a bare proton – can attach to an electron pair, and hence is an acid.

For the purpose of this discussion we need to note how the characteristic of acid–base behaviour has migrated away from the particle being transferred to the properties of electrons and their ability to participate in the bond formation. In a somewhat tongue-in-cheek way, I have suggested that one can broaden the definition even further, and define a base as an electron and an acid as a hole in which the electron can fit. Then all the Lewis chemistry, Brønsted–Lowry chemistry, and Arrhenius chemistry survive, and *all* chemical reactions (including oxidations and reductions) are seen to be aspects of acid--base behaviour!

The pragmatic usefulness of the various explanations of acid–base behaviour declines somewhat with their generality. Chemists still sometimes think in Arrhenius terms if they are dealing with aqueous solutions. The Brønsted–Lowry definitions represent an excellent balance between generality and focus on a specific entity, the hydrogen ion, and are widely used on that account. Although the Lewis explanation is even more general, it is almost too general, and chemists invoke it in the main when they are dealing with reactions that involve species other than hydrogen ions. The frontier between these two approaches also represents the type of calculations that can be done, for the body of calculations that can be carried out within the context of the Brønsted–Lowry definitions does not really survive when extended to the whole domain of Lewis chemistry. That is also true in even greater measure of the extension of the definition to electrons and holes: although conceptually it may be pleasing to see all chemical reactions as having a common explanation, in practice different classes of reactions have their own apparatus of calculation and insight, and there appear to be no advantages in trying to unify them into a single corpus of discussion. Breadth of explanation is often best kept as the backdrop of understanding.

These reflections on the nature of chemical reactions lead naturally to a consideration of why any reaction takes place at all: what is the motive power of change in chemistry? The explanation of chemical change emerged in the nineteenth century and the formulation of the laws of thermodynamics. In particular the Second Law of thermodynamics, that the entropy of an isolated system never decreases, is crucial to our understanding of why anything happens at all. Although thermodynamics emerged from a study of steam engines, it was brought into chemistry largely through the careful thoughts of J. W.

Gibbs, and enriched by the molecular understanding of entropy that was formulated by Ludwig Boltzmann. Now we know that the motivation of all chemical change (and therefore of all biological change too) is collapse into disorder: energy and matter tend to disperse in disorder. That single idea is sufficient to explain every chemical reaction. Moreover, through the subtle mathematics of chemical thermodynamics, chemists are able to predict the spontaneous direction of chemical reactions (whether a reaction will tend to run in one direction or another) and the position of equilibrium, when a reaction has no tendency to run in either direction.

Classical thermodynamics is a corpus of relations between observations on bulk systems, which allows measurements of one property to be used to predict another. Statistical thermodynamics, the formulation of thermodynamics in terms of the behaviour of the myriads of molecules that make up bulk matter, is an enrichment of classical thermodynamics and provides profound insight into the events that go on in bulk matter and which lead to reaction. Statistical thermodynamics is also extraordinarily useful on account of its forming the bridge between spectroscopic determinations of molecular properties and thermodynamic descriptions of bulk properties.

The point should not be missed that thermodynamics, and therefore the explanation of the direction of chemical reaction, emerged from studies of the steam engine. At first sight, a steam engine seems far removed from the delicate events that constitute a chemical reaction in a test-tube or the unfolding of a leaf. But one of the glories of science is that the abstract essence of one phenomenon will be found in another. In this case, the abstract essence of a steam engine is found in every chemical reaction, including those in industry, in laboratories, and in biology. The essence of a steam engine is the removal of energy stored at a high temperature and its deposit in a sink maintained at a lower temperature. The first stage results in a decrease in entropy; the second results in a greater increase in entropy. Thus, the dispersal of energy is used to drive whatever the engine is connected to. The same is true in chemistry, where one reaction, such as the metabolism of a food, can be used to drive another to which it is connected, such as the formation of proteins mediated by enzymes. Whenever you look at a process in the world, you should view it as the realization of the abstraction of a steam engine.

Thermodynamics is an explanation of the natural direction of chemical change. It predicts the tendency to change. It is silent on the rate at which that natural change might come about. The formation of water from hydrogen and oxygen is a spontaneous change in the technical sense of having a tendency to occur, but a mixture of the two gases can survive indefinitely until a spark sends the reaction on its way. Chemists are careful to distinguish the tendency to change from its rate, the latter being the domain of chemical kinetics.

Chemical kinetics emerged towards the end of the nineteenth century when chemists first collaborated with mathematicians and discovered that simple differential equations could be applied to chemical change to predict the composition of reaction mixtures as a function of time. There soon emerged an understanding of why reactions proceeded at different rates and how those rates depended on the temperature: the explanation was that there is a barrier to reaction, and only those species encountering each other with at least a certain energy could cross the barrier.

Two central developments have brought chemical kinetics to its current standing. The differential equations that describe the evolution of chemical reactions are strikingly difficult: because a chemical reaction may proceed in a sequence of steps with positive and negative feedback, in the sense that a product formed in one step may be the reactant in an earlier step, the equations are highly non-linear. Until the advent of computers, chemists had no option but to simplify these equations by forcing them to be simple algebraic equations, thereby losing almost all their richness. Computers brought numerical techniques to bear, and now the full grandeur of the rate equations can be unfolded. Now chemists can deal with chemical reactions that give rise to awesome spatial patterns that evolve in time. A by-product of this achievement is that we can now explain the patterns of animal pelts, for a pelt is just an oddly shaped test-tube, and the rhythm of the heartbeat.

The second development in chemical kinetics that has brought us close to full understanding of chemical reactions is the ability to monitor encounters between individual molecules. Through the agency of molecular beams – which are streams of molecules that are allowed to collide under carefully controlled conditions – and lasers, which give information about chemical reactions on a timescale of femtoseconds (1 fs = 10^{-15} s), we can now monitor the most intimate moments of reactions.

This discussion of explanation in chemistry began with the periodic table, and it is appropriate to conclude by looking at the distribution of different types of explanation across that table. Broadly speaking, organic chemistry – the chemistry of carbon – invokes a combination of explanations in terms of structure and kinetics. Organic compounds have an exquisite architecture, and molecular architecture to a large degree determines function. That is particularly the case when chemists turn their attention to protein function. Although life has to be understood in the light of the Second Law as the great unwinding of the cosmos, we can understand its richness by recognizing that there are profound kinetic constraints on the rates at which reactions take place. Life would not be life if it were over in a flash. Life is life because of the complexity of the network of reactions that mediate and retard the Great Unwinding.

Elsewhere in the periodic table, throughout the domain of inorganic chemistry, structure and thermodynamics are the two principal modes of explanation. Many of the most important compounds that inorganic chemists now strive to understand are solids, and to understand solids it is essential to understand their structure. Many of the reactions that inorganic chemists consider are controlled more by thermodynamics and suffer less kinetic inhibition than is characteristic of the compounds of carbon. Inorganic chemistry also invokes portmanteau properties, such as electronegativity and chemical hardness, which are ill-defined composites of more fundamental properties, and yet which put considerable rationalizing power into the hands of the circumspect chemist.

But none of these generalities is exclusive, and the richness of chemistry stems from the interplay of different explanations that emphasize a particular aspect of the problem in a manner that makes good sense. In short, explanation in chemistry is more rationalizing than predictive. The subject lies on the frontier between simplicity and complexity, and the mode of explanation varies with the type of property being considered. Explanation in chemistry is an intricate network of components that illuminate different facets of an intricate subject.

Steven Rose

7

The biology of the future
and the future of biology[1]

Constraints on knowing nature

WE view and interpret the world around us, both natural and cultural, through perceptual and cognitive spectacles of our own construction. The natural sciences claim that their methods, of hypothesis, observation, and experiment, permit something approximating to a true representation of the material reality that surrounds us to be achieved. However, for several decades now, philosophers, historians, and sociologists of science have been pointing to the ways in which our scientific knowledge is socially, culturally, and historically constructed; that is, it offers at best a constrained interpretation of the material world. A first such constraint is provided by the very construction of our brain and the biology of our perceptual processes. To this I would add that brains do not exist in isolation from bodies: how we perceive the world is affected by our hormonal, immunological, and general physiological state. And we perceive the world in the way that we do because our visual system is capable of sensing only a limited range of wavelengths, our mass and volume give us a particular relationship to gravitational forces not shared for instance by bacteria or beetles, or by whales or elephants. Our sense of the temporality of events is shaped by the fact that we may live for anything up to and now even beyond a century. Bacteria divide every twenty minutes or so, mayflies live for a day, redwood trees for thousands of years. Human technologies can and do enable us to escape these structural and temporal limitations, to observe in the infrared or ultraviolet, weigh atoms, and measure time in anything from nano-seconds to light years. Yet even when considering the inconceivably small or distant, we do so by scales that relate to our human con-dition: the measure of man is man.

But there are other constraints that transcend our mere biology. One – which is where the sciences differ most from the arts and humanities – is that we are not free to offer interpretations which our observations of, and experiments on, the external world disconfirm. A second is defined by the limits of our available technologies. Until the means of circumnavigating the earth were available it was a legit-imate approximation to the truth to maintain that the earth was flat. Until Lavoisier weighted the products of combustion, phlogiston theory was as good as oxygen theory. Until microscopes revealed the internal constituents of living tissue it was legitimate to regard cells as composed of homogeneous protoplasm.

But a third and equally important constraint is that resulting from the very social and historical nature of the scientific enterprise itself. The ways in which we view the world, the types of experiments we conceive and evidence we accept, the theories we construct, are far from being culturally free. This means that we cannot understand the current shape of biological thinking without reference to the history of our own discipline. This should not surprise us. After all, in a famous aphorism, known to all biologists, the great evolutionist Theodosius Dobzhansky pointed out that nothing in biology made sense except in the context of evolution. I would want merely to broaden that statement, in ways that will become apparent in the course of this paper, to read 'nothing in biology makes sense except in the context of history' – by which I mean evolutionary history, devel-opmental history, and the history of our own subject. This does not imply a simple progressivism; new knowledge claims may well be, but are not necessarily, 'better' representations of the material world than prior ones.

The power of reductionist thinking and the plurality of biological explanation

For reasons that it would take me too far outside my theme to explore here, throughout its post-Cartesian and Newtonian history Western science has seen physics as its explanatory model. The more pluralistic, pre-scientific world gave way to one in which all our day-to-day experiential richness of colour, and sound, of love and anger, came to be seen as secondary qualities underlying which there were the changeless particles, waves, and forces of the physicists' world. The task of other sciences, chemistry, biology and later psychology, sociology, and economics, was to make themselves as like physics as possible. Their qualities needed to be reduced to the true, quantitative, 'hard science' of physics. As philosopher Thomas Nagel claims, other sciences describe things, reductionism explains them. The molecular biologist, James Watson, following Rutherford, put it more bluntly: 'there is only one science, physics; everything else is social work.'

A further problem for biology, perhaps because its subject matter is so complex, is that there has been a continuing tendency to understand living processes and systems by metaphorizing them to the most advanced forms of current human artefact. Many origin myths, including the Judaeo-Christian one, refer to humankind having been created from the dust or clay of the earth, as on a potter's wheel. The Greek Hero was said to have created lifelike robots using steam and water-power. Hydraulic imagery persisted through the Renaissance (for example hearts as pumps; nerves as tubes for conveying vital forces) to give way to electrical and magnetic ones in the eighteenth and nineteenth centuries. Brains became telegraph systems, then telephones, and now computers. Such metaphors are powerful, and may be helpful. But too often their seductive powers blinker our capacity to see the world. As I will argue, brains are not computers, and genes are not selfish.

Let me offer a fable to demonstrate the limits of reductionism in biology. Five biologists are on a picnic, when they see a frog jump into a pond. They fall to discussing why the frog has jumped. The first, a physiologist, describes the frog's leg muscles and nervous system. The frog jumps because impulses have travelled from the frog's retina to its brain and thence down motor nerves to the muscle. The second, a biochemist, points out that the muscles are composed of actin and myosin proteins – the frog jumps because of the properties of these fibrous proteins that enable them, driven by the energy of ATP, to slide past one another. The third, a developmental biologist, describes the ontogenetic processes whereby the fertilized ovum divides, in due course forming the nervous system and musculature. The fourth, a student of animal behaviour, points to a snake in a tree above where the frog was sitting: the frog jumps to escape the snake. The fifth, an evolutionist, explains the processes of natural selection that ensured that only those frog ancestors able both to detect snakes and jump fast enough to escape them had a chance to survive and breed.

Five biologists, five very different types of explanation. Which is the right one? Answer: all of them are right, just different. The biochemist's explanation is the reductionist one, but it in no way eliminates the need for the others. Nor can we envisage a research programme in which the other types of explanation would in due course all be subsumed by either the biochemical, or the evolutionary one – despite the often quoted claim in the opening sentences of E. O. Wilson's book on sociobiology, reiterated most recently in his demand for 'consilience'. The most that we can insist is that explanations in the different discourses should not contradict one another. As a materialist, as all biologists must be, I am committed to the view that we live in a world that is an ontological unity, but I must also accept an epistemological pluralism. As the philosopher Mary Midgley puts it, neither the value of money, nor the rules of football, are collapsible into physics; there is one world, but it is a big one.

The shifting boundaries between nature and culture

Which type of explanation we prefer depends on the purposes for which it is required. If we are concerned with diagnosing and treating diseases like muscular dystrophy or myasthenia gravis, genetic and biochemical approaches may point the way. For others, like understanding why a person chooses to take a swim or a baby learns to walk, they are almost entirely useless. Far from such 'lower level' accounts explaining, as Nagel would have it, they at best merely describe, whilst the higher and strictly irreducible accounts have the most explanatory power. Yet over the hundred and forty years since the publication of Darwin's *Origin*, biology's pluralism has steadily been restricted. Physiology has been collapsed into biochemistry and biochemistry into chemistry and physics. Skinner's behaviouristic attempt to reduce psychology to physics and skip the intervening biological level may have been rejected, but a new school of 'neuro-philosophers' seeks to dismiss what they call 'folk psychology' in favour of neurocomputation, in which brains are indeed replaced by computers. Evolutionary biology has replaced study of the living world by abstract mathematical calculations about changes in the population frequencies of individual selfish genes. Indeed, the very phenotype, the living organism itself, has been emptied of any function other than that of being the 'lumbering robot' serving for the replication of its genes, to quote Dawkins. Molecular geneticists now see organisms as mere tools with which to probe gene function. They offer to predict our entire lifeline, our trajectory from birth to death, from the diseases we will die of to the political parties we will vote for, the drugs we will enjoy, our degree of job satisfaction, and our tendency to midlife divorce. It would seem that reductionism has triumphed.

This reconceptualization of what is 'nature' is moving forward with great speed on the heels of new developments in genetics. Thus the neurosciences in collaboration with genetics are offering to transform our understanding of human nature, turning what were once regarded as the results either of individual free will (or humanity's sinful nature – see for instance Milton's *Paradise Lost* for an earlier orthodox theological account) into biomedical matters. Adultery and cheating were once sins; now evolutionary psychology tells us that these activities are the consequences of adaptations during humanity's palaeolithic past. Alcoholism and violence were social problems; now they are supposed to be caused by abnormal genes. Same-sex love was once a sin, then a social problem; today there are claims that it is genetically caused. Personal responsibility for our actions is dissolved into DNA's double helix. Culture becomes nature, whilst simultaneously biomedicine holds out the promise of transforming nature by genetic manipulation. Biology is Janus-faced. It is determinist, as much predestinationist as any religious sect. But it is simultaneously Promethean, offering technology to conquer destiny.

Yet there are limits to reductionism's onward march. Try as we may to collapse our sciences into grand physical theories of everything (the physicists' TOES and GUTS), such attempts must fail. The dream of writing a single equation which will embrace the world resembles more the search for an alchemical philosopher's stone which will transmute base metal into gold and grant perpetual youthful life to its possessor, or the cabbalistic belief that there can be found a single mystical sentence which will give its utterer almost god-like power over people and things. The Vienna School's philosophical attempt in the 1930s to impose a unity on the sciences – a unity built upon physics – was doomed to failure, and Wilson's latest evocation of it is unlikely to succeed.

Let me draw briefly on three areas of current biology to demonstrate the limits to reductionism's dream and use these to point in the directions in which I believe biology must move in its concept of nature in the new millennium: neuroscience, developmental biology, and evolutionary biology.

Brains, minds, and meaning

First, the brain. To understand how the human brain functions it is not sufficient to simply extrapolate upward from the firing properties of its hundred billion neurones or the 10^{15} or so synaptic connections between them. Even a wiring diagram of their anatomical connections is insufficient. The temporal relationships between activity in distant brain regions, coherent oscillatory processes, binding mechanisms, and doubtless as-yet undiscovered interactions all make it necessary to consider the brain as a system, not a mere assemblage of parts, perhaps to be understood using chaotic dynamic theory. I want to add one further point, and that is to challenge the popular view that brains are computational, information processing devices, a claim most forcefully made recently by Steven Pinker.

In Pinker's view, following other evolutionary psychologists, the brain/mind is not a general-purpose computer; rather it is composed of a number of specific modules (for instance, a speech module, a number sense module, a face-recognition module, a cheat-detector module, and so forth). These modules have, it is argued, evolved quasi-independently during the evolution of early humanity, and have persisted unmodified throughout historical time, underlying the proximal mechanisms that traditional psychology describes in terms of motivation drive, attention, and so forth. Whether such modules are more than theoretical entities is unclear, at least to neuroscientists. Indeed evolutionary psychology theorists such as Pinker go to some lengths to make it clear that the 'mental modules' they invent do not, or at least do not necessarily, map onto specific brain structures. But even if mental modules do exist they can as well be acquired as innate.

Modules or not, it is not adequate to reduce the mind/brain to nothing more than a cognitive, 'architectural' information processing machine. Brains/minds do not just deal with information. They are concerned with living meaning. In *How the mind works*, Pinker offers the example of a footprint as conveying information. My response is to imagine Robinson Crusoe on his island, finding a footprint in the sand. First he has to interpret the mark in the sand as that of a foot, and recognize that it is not his own. But what does it mean to him? Pleasure at the prospect of at least another human being to talk and interact with? Fear that this human may be dangerous? Memories of the social life of which he has been deprived for many years? A turmoil of thoughts and emotions within which the visual information conveyed by the footprint is embedded. The key here is emotion, for the key feature which distinguishes brains/minds from computers is their/our capacity to experience emotion. Indeed, emotion is primary; affect as much as cognition is inextricably engaged in all brain and mind processes, creating meaning out of information – just one reason why brains aren't computers. What is particularly egregious in this context is Pinker's oft-repeated phrase, 'the architecture of the mind'. Architecture, which implies static structure, built to blueprints and thereafter stable, could not be a more inappropriate way to view the fluid dynamic processes whereby our minds/brain develop and create order out of the blooming, buzzing confusion of the world which confronts us moment by moment.

DNA and the cellular orchestra

Within the reductionist paradigm, development is the reading out of genetically encoded instructions, present within the 'master molecule', DNA. Hence the popular references to DNA as being the blueprint of life, the code of codes, and so forth. The organism, the phenotype, is simply the vessel constructed by the DNA in order to ensure its safe replication. This model misspeaks both the relationship of DNA to cellular processes in general and the nature of development.

DNA itself is rather an inert molecule (hence the possibility of recovering it intact from amber many tens of thousands of years old – and the plot of *Jurassic Park*). What brings it to life is the cell in which it is embedded. DNA cannot simply and unaided make copies of itself; it cannot therefore 'replicate' in the sense that this term is usually understood. Replication – using one strand of the double helix of DNA to provide the template on which another can be constructed – required an appropriately protected environment, the presence of a wide variety of complex molecular precursors, a set of protein enzymes, and a supply of chemical energy. And even when DNA has been copied faithfully, the 'read-out' into RNA and thence into protein, once thought to be linear, is far from being so. Individual coding sequences of DNA are distributed along chromosomes, punctuated by long sequences (introns) of no known coding function. In humans some 98% of the DNA in the genome comes into this non-coding category. Coding sequences are 'read' by cellular mechanisms, snipped out from the rest, spliced together, transcribed into RNA, edited.

The proteins they code for are then further processed and tailored by temporal and state regulated cellular mechanisms quite distal to DNA itself. All these are provided in the complex metabolic web within which the myriad biochemical and biophysical interactions occurring in each individual cell are located. The famous 'central dogma' enunciated by Crick and on which generations of biologists have been brought up, that there is a one-way flow of information by which 'DNA makes RNA makes protein' and that 'once information gets into the protein it can't flow back again', was a superb simplification in the early days of molecular biology. But it simply isn't true any more. Dogma in science is as unstable as that in religion. What is certain is that there are no master molecules in cellular processes. Even the metaphor of the cellular orchestra, which I have used previously, is not adequate, as orchestras require conductors. Better to see cells as marvellously complex versions of string quartets or jazz groups, whose harmonies arise in a self-organized way through mutual interactions. This is why the answer to the chicken and egg question in the origin of life is not that life began with DNA and RNA but that it must have begun with primitive cells which provided the environment within which nucleic acids could be synthesized and serve as copying templates.

The paradox of development

It is a commonplace that, despite the near identity of our genes, no-one would mistake a human for a chimpanzee. For that matter we share some 35% of our genes with daffodils. What distinguishes even closely related species are the developmental processes that build on the genes, the ontogenetic mechanisms that transform the single fused cell of a fertilized ovum into the thousand trillion cells of the human body, hierarchically and functionally organized into tissues and organs. Developmental processes have trajectories which constitute the individual lifeline of any organism, trajectories which are neither instructed by the genes, nor selected by the environment, but constructed by the organism out of the raw materials provided by both genes and environment.

One of the problems for twentieth century biology, a problem resulting partly from contingent features of the history of our science, was that whereas at the beginning of the century developmental biology and genetics were seen as a single scientific endeavour (a splendid exemplar being T. H. Morgan, the founder of the famous 'fly school' which introduced *Drosophila* as 'God's organism' for genetic and chromosomal analysis), by the 1930s they had become quite distinct. Thus development became the science of similarities, genetics the science of differences. Explaining how it is that virtually all humans are between 1.5 and 2.5 metres in height, are more or less bilaterally symmetrical, and have pentadactyl limbs was a subject for developmental study. Why some of us have differently coloured eyes, hair, or skin became part of genetics. Only in the closing years of the century did there seem to be a chance to bring the two together once more.

The unity of an organism is a process unity, not a structural one. All its molecules, and virtually all its cells, are continuously being transformed in a cycle of life and death which goes on from the moment of conception until the final death of the organism as a whole. This means that living systems are open, never in thermodynamic equilibrium, and constantly choosing, absorbing, and transforming their environment. They are in constant flux, always at the same time both being and becoming. To build on an example I owe to Pat Bateson, a newborn infant has a suckling reflex; within a matter of months the developing infant begins to chew her food. Chewing is not simply a modified form of suckling, but involves different sets of muscles and physiological mechanisms. The paradox of development is that a baby has to be at the same time a competent suckler, and to transform herself into a competent chewer. To be, therefore, and to become.

Being and becoming are not to be partitioned into that tired dichotomy of nature versus nurture. Rather they are defined by a different dichotomy, that of specificity and plasticity. As an example, consider the relationship between eye and brain. The retina of the eye is connected via a series of neural staging posts to the visual cortex at the back of the brain. A baby is born with most of these connections in place, but during the first years of life the eye and the brain both grow, at different rates. This means that the connections between eye and brain have continually to be broken and remade. If the developing child is to be able to retain a coherent visual perception of the world this breaking and remaking must be orderly and relatively unmodifiable by experience. This is specificity. However, as both laboratory experiments and our own human experience show, both the fine details of the 'wiring' of the visual cortex and how and what we perceive of the world are directly and subtly shaped by early experience. This is plasticity. All living organisms and perhaps especially humans with our large brains show both specificity and plasticity in development, and both properties are enabled by our genes and shaped by our experience and contingency. Neither genes nor environment are in this sense determinant of normal development; they are the raw materials out of which we construct ourselves.

Thus the four dimensions of living processes – three of space and one of time – cannot be read off from the one-dimensional strand of DNA. A living organism is an active player in its own destiny, not a lumbering robot responding to genetic imperatives whilst passively waiting to discover whether it has passed what Darwin described as the continuous scrutiny of natural selection.

Evolution and levels of selection

Within the reductionist paradigm within which much of contemporary biological theorizing is located, the processes of evolutionary change have also become simplified. By contrast, for instance, with Darwin's own pluralism, which saw natural selection as a main but not the only mechanism of change, a dominant orthodoxy, described as fundamentalist or ultra-Darwinian, has emerged. Three main theses characterize this new fundamentalism. First, most phenotypic features we can observe are adaptive; second, they are generated by natural selection; and third, natural selection acts solely or primarily at the level of individual genes. A new biology for the new millennium must transcend each of these propositions, in part by reverting to Darwin's own more pluralistic understanding. I will consider the counter-arguments to these propositions in reverse order.

The present-day understanding of the fluid genome in which segments of DNA responsible for coding for subsections of proteins, or for regulating these gene functions, are distributed across many regions of the chromosomes in which the DNA is embedded, and are not fixed in any one location but may be mobile, makes the view that individual genes are the only level of selection untenable. But it always was. To play their part in the creation of a functioning organism many genes are involved – in the human, some thirty thousand. For the organism to survive and replicate, the genes are required to work in concert – that is, to co-operate. Antelopes which can outrun lions are more likely to survive and breed than those that cannot. Therefore a mutation in a gene which improves muscle efficacy, for instance, might be regarded as fitter and therefore likely to spread in the population. However, as enhanced muscle use requires other physiological adaptations – such as increased blood flow to the muscles – without this concerted change in other genes, the individual mutation is scarcely likely to prove very advantageous. And as many genes have multiple phenotypic effects (pleiotropy) the likelihood of a unidirectional phenotypic change is complex – increased muscle efficacy might diminish the longevity of the heart, for example. Thus it is not just single genes which get selected, but also genomes. Selection operates at the level of gene, genome, and organism.

Nor does it stop there. Organisms exist in populations (groups, demes). Three decades ago, Wynne-Edwards argued that selection occurred at the level of the group as well as the individual. He based this claim on a study of a breeding population of red grouse on Scottish moors, and argued that they distributed themselves across the moor, and regulated their breeding practices in a way which was optimum for the group as a whole rather than any individual member within it. It may be in the individual's interest to produce lots of offspring; but this might overcrowd the moor, which could only sustain a smaller number of birds; hence it is in the group's interest that none of its members overbreed. Orthodox Darwinians, led by George Williams, treated this claim with as much derision as they did Lamarck's view that acquired characteristics could be inherited, and group selection disappeared from the literature.

Today, however, it is clear that the attack was misjudged. To an extent it was always in part semantic. Maynard Smith's work, for instance, indicated that stable populations require the mutual inter-actions and ratios of members with very different types of behaviours (hawks and doves, for instance, to use his model) – so-called evo-lutionary stable strategies. But there are an increasing number of examples of populations of organisms whose behaviours can most economically be described by group selectionist equations. Recently Sober and D. S. Wilson have published a major reassessment of group selectionist models and shown mathematically how even such famously counter-intuitive (for ultra-Darwinians) phenomena as altruism can occur, in which an individual sacrifices its own individ-ual fitness, not merely for the inclusive fitness of its kin but for the benefit of the group as a whole.

Finally, there is selection at even higher levels – that of the species, for example. Natural selection may be constantly scrutinizing and honing the adaptiveness of a particular species to its environment, but it cannot predict the consequences of dramatic changes in that environment, as for example the meteor crash into the Yucatan believed to have precipitated the demise of the dinosaurs. Selection also operates at the level of entire ecosystems. Consider, for example, a beaver dam. Dawkins uses this example to claim that the dam may be regarded as part of the beaver's phenotype – thus swallowing an entire small universe into the single strand of DNA. But if it is a phenotype, it is the phenotype of many beavers working in concert, and indeed of the many commensal and symbiotic organisms which also live on and modify for their needs the structure of the dam. As Sober and Wilson point out, selection may indeed occur at the level of the individual, but what constitutes an individual is very much in the eye of the beholder. Genes are distributed across genomes within a population. There is no overriding reason why we should consider 'the organism' as an individual rather than 'the group' or even 'the ecosystem'.

Nor is natural selection the only mode of evolutionary change. We need not be Lamarckian to accept that other processes are at work. Sexual selection is one well-accepted mechanism. The existence of neutral mutations, founder effects, exaptations, genetic drift, and molecular drive all enrich the picture. Gould argues that much evolutionary change is contingent, accidental, and that, as he puts it, if one were to wind the tape of history backward and replay it, it is in the highest degree unlikely that mammals, let alone humans, would evolve.

Finally, not all phenotypic characters are adaptive. A core assumption of ultra-Darwinism is that observed characters must be adaptive, so as to provide the phenotypic material upon which natural selection can act. However, what constitutes a character – and what constitutes an adaptation – is as much in the eye of the beholder as in the organism to which the 'character' belongs. The problem lies in part in the ambiguity of the term phenotype, which can refer to anything from a piece of DNA (strictly the gene's phenotype) through the cellular expression of a protein to a property of the organism as a whole, like height, or a 'behaviour', such as gait. At which level of phenotype a character is 'adaptive', if at all, and at which its properties are epiphenomenal, is always going to be a matter for debate. A not entirely apocryphal example is provided by the American artist Thayer who suggested, early last century, that the flamingos' pink colouration is an adaptation to make them less visible to predators against the pink evening sky. But the colouration is a consequence of the flamingo's shrimp diet and fades if the diet changes. Thus even if we were to assume that the colouration was indeed protective, it is an epiphenomenal consequence of a physiological – dietary – adaptation, rather than a selected property in its own right.

Natural selection's continual scrutiny does not give it an à la carte freedom to accept or even reject genotypic or phenotypic variation. Structural constraints insist that evolutionary, genetic mechanisms are not infinitely flexible but must work within the limits of what is physically or chemically possible. For instance, the limits to the size of a single cell are set by the physics of diffusion processes, and the size of a crustacean like a lobster or crab by the constraints of its exoskeleton. The limits to the possible lift of any conceivable wing structure make it impossible to genetically engineer humans to sprout wings and fly; there are good reasons why we cannot become angels.

A biological decalogue for the millennium

If we are to transcend biology's reductionist view of nature for the new millennium, and to create what I would regard as an understanding of living processes more in accord with the material reality of the world than our present, rather one-eyed view, we need some principles with which to work. These principles will, of course, not reject the explanatory power of reductionism, but will recognize its limitations. They should accept, too, Goodwin's call for the return to a science of qualities, to complement reductionist quantitation. Such a science will rejoice in complexity, in dynamics, and in an emphasis as much on process as on objects. In my book *Lifelines* I enumerate a set of such principles; by chance rather than design, there are ten in all. In brief, here is my Decalogue. To exemplify all its principles would extend this essay beyond reasonable length, but enough has I trust been said to give a feel for the conceptual approach I am arguing for:

- Scientific knowledge is not absolute, but provisional, being socially, culturally, technologically, and historically constrained.

- We live in a world that is ontologically unitary, but our knowledge of it is epistemologically diverse. There are multiple legitimate ways of describing and explaining any living process.

- Different sciences deal with different levels of organization of matter of increasing complexity. Terms and concepts applicable at one level are not necessarily applicable at others. Thus genes cannot be selfish; it is people, not neurones nor yet brains or mind, who think, remember, and show emotion.

- Causes are multiple and phenomena richly interconnected. Adequate explanation demands finding the determining level. To take an example, the high levels of murder in the US, by comparison with, say, Europe or Japan, are best explained not by some special feature of the US genotype, abnormal genes, or biochemistry which predispose to violence (despite a major research programme dedicated to identifying such biological predispositions), but by the high number of personal handguns in society, and a culture and history of their use.

- Living organisms exist in four dimensions, three of space and one of time, a developmental trajectory of lifeline, always autopoietic, both being and becoming. Lifelines are stabilized through dynamic processes. The traditional biological concept of homeostasis as a regulatory mechanism needs transforming by that of homeodynamics, to emphasize that indeed stasis for any living organism means death.

- Organism and environment interpenetrate; environments select organisms (the process of natural selection); but organisms choose and transform environments. Organisms are thus active players in their own destiny.

- Living organisms are open systems far from thermodynamic equilibrium; continuity is maintained by a constant flow of energy and information. All is flux; stability emerges through process, not structure.

- Evolutionary change occurs at the intersection of lifeline trajectories with changing environments.

- Organisms cannot predict patterns of change, and selection therefore always tracks environmental change. Nothing in biology makes sense except in the context of history.

- Thus the future, for humans and other living organisms, is radically unpredictable; we make our own history, though in circumstances not of our own choosing.

Note

1. This is an edited version of a paper that first appeared in *Perspectives in Biology and Medicine*, 44(4), 473–84, The Johns Hopkins University Press, 2002. Reprinted by permission.

David Hanke

8

Teleology:
the explanation that bedevils biology

BIOLOGY is sick. Fundamentally unscientific modes of thought are increasingly accepted, and dominate the way the subject is explained to the next generation. The heart of the problem is that we persist in making (literally) sense of a world that we now know to be senseless by attributing subjective values to the objects in it, values that have no basis in reality.

For examples, read anything written by the first Oxford Professor of Public Understanding of Science, probably the most influential communicator of biology in his generation. Here's a smattering from the book he intended for professional biologists:

> Somatic cell divisions are used to make mortal tissues, organs and instruments whose 'purpose' is the promoting of germ line divisions.

> Only the germ-line cells, it would seem, really need to preserve the entire genome. It may be that the reason is simply that there is no easy way, physically, to hive off parts of the genome.

> The organism is a unit with a life cycle which, however complicated it may be, repeats the essential characteristics of previous life cycles, and may be an improvement on previous life cycles.

The behaviour of an individual may not always be inter-
pretable as designed to maximise its own genetic welfare.

The replicators that exist tend to be the ones that are good at
manipulating the world to their own advantage.

(Dawkins 1982)

Professor Dawkins' more obvious 'metaphors of purpose', for example 'DNA is not working for the good of the cell but for the good of
itself' (and the fantasy chat between genes on p. 212) possibly are as
he claims 'harmless anthropomorphisms', lively literary devices to
engage the reader. Occasionally he uses quotes to flag up metaphorical use, as for 'purpose' above, but 'need', 'easy', 'improvement',
'designed', 'welfare', 'good', and 'advantage' are subjective value
judgements that slip in under the radar. It is entirely admirable that
every candidate for admission to biological sciences at Cambridge has
read at least one of Professor Dawkins' books. The tragedy is that
very few of them have read anything else.

Here are some quotes from a recent review of the domain structure of the photosynthetic membrane, not a subject for whose
explanation metaphors of purpose should be necessary, one feels.

Choroplasts have been evolutionarily successful.

Hence, it is no surprise that the photosynthetic membrane has
an ingenious structure to fulfil all its functions.

This is one reason why nature developed macromolecules.

Randomness had to be avoided.

(Albertsson 2001)

When I was taught the core principles of scientific method, nearly half
a century ago, I was trained to reason critically and objectively, with

the aim of uncovering the reality out there, and to regard the subjective as untrustworthy and deceiving, potentially corrupting the
truth. One had to work at weeding it out by rejecting value judgements, opinions – in fact any idea not securely backed up by factual
evidence. It was necessary to be constantly on one's guard against
a form of subjective thinking called 'teleology' to which biology is
especially susceptible. Teleology is the mental habit of supposing that
objects or events have a purpose. All human activities are intensely
purposive, we do things for a purpose, we make things for a purpose.
Our lives and thoughts are drenched in purposes, our own and those
of other people. It seems to be very hard to climb outside this distorting mindset. If I was an evolutionary biologist (which I'm not) I'd be
tempted to speculate that the human propensity to find purpose in
things and events evolved because it contributed to the survival of
the individual beset by purposeful predators and potentially ill-
intentioned competitors. It's not difficult to understand why teleology
is tempting for observers trying to explain the world of living things.
All the individuals in it are festooned with business-like widgets.
Our first question is, naturally enough, 'What's that *for*?' The sheer
efficiency of biological structures reinforces the illusion of purpose.

The purpose of any object is entirely subjective because purpose
has no real existence outside the mind of the animal thinking of it.
Objects intended for use in some future event, a sling-shot stone or a
cog for a gearbox, say, can correctly be said to have a purpose. Their
intended use can, at least in theory, be known – although only from
the intending user, not from the object itself. In the absence of the user,
we can only guess an intended use from clues offered by the object
itself. The gearbox cog provides evidence of its purpose, the sling-shot
stone none at all. The cog is symmetrical, precise, and complex, properties that suggest to us that it is the end product of a sequence of conscious design guiding a subsequent process of manufacture.

The bits of living things at all levels of scale from molecules to the
whale's tail also happen to have symmetry, precision, and complexity,

clues that simply shout 'purpose' in the enquiring mind. This has to be wrong. Because they are not manufactured they cannot have been designed, and so no-one ever had a purpose for them. They make themselves and so just exist, without purpose or intended use.

There is a class of objects within the natural world that differs from the bird's wing, the beaver's tail, and the ant's jaws in that they have been truly built. Does that make it legitimate to suppose a purpose for nests and dams? Lacking Dr Doolittle's powers, we can't know the intentions of the animals as they make these things. We don't know the extent to which they are aware of a purpose for what they build. Given the interrelatedness of all living things, it seems churlish to deny, at least for our nearest relatives, a degree of awareness of the future consequences of their actions – an idea of why they are constructing the artefact, or the uses to which it might be put, but, it would be argued, theirs will be a dimmer intention as befits a so-called 'lesser' intelligence. It's hard to see how beavers can accommodate the uniqueness of each new site unless they have some mental vision of the end result. On the other hand, it's equally hard to see how the complete plan of an ants' nest in its entirety can ever have existed inside the head of any individual ants because we cannot conceive of a mechanism by which the apparently random teeming of workers could convert any complex mental goal into reality. We're happier with the idea that the form of the nest is an emergent property arising as a large-scale consequence of the actions of a number of individuals responding stereotypically to their small-scale surroundings. So, to conclude, we can only be sure of the purpose of a built object, in the original sense of the builder's intended use, if the builder is human and communicative. If not, we're relying on supposition.

On top of all this, as humans we find it hard to resist imposing our own ideas about the purpose of animal constructions quite independently of the intentions of the builder. Sometimes this extra layer is referred to as the 'true' or 'higher' purpose, and for a built object a comparison with the actual purpose (the intentions of the builder) illuminates the illogicality of the teleological approach. As creatures who make things for specific purposes ourselves, we are sure we know what the thrushes nest is for, even if the thrush herself didn't have a clue why she followed a stereotypical sequence of mechanical actions, driven by a mindless urge to wattle and daub. We decided what the 'real' purpose is by imagining what we would need if we were birds; another type of subjective reasoning, this time 'anthropomorphism'. Use of the word 'need' is a reliable indicator of underlying anthropomorphic thinking. The 'needs' of plants are frequently given solemn consideration, the writer usually constructing a helpful shopping list for those benighted creatures. The depth of condescension plumbed can be felt more personally by reflecting on the affront engendered by the trivialization of human sexual relations when evolutionary biologists pontificate about its 'true' purposes. The fact is, unlike the intentions of the nest-building bird, which are at least real even if currently unknowable, any purpose we might attach to the nest is no more real than our speculations about what a thrushes wing is for. Our idea of its purpose exists only in our heads.

It is no longer acceptable to think of biological objects as having any purpose because the overwhelming consensus of scientific opinion is that they were not designed and built by a Creator (a mental construct necessary to inject a human sense of purpose into existence) with purposes in mind for them. Instead, we believe (I'll put that as strongly as I can) they are products of Darwinian evolution. Now this ought to mean that we can turn with some relief to the scientists who study the mechanism of evolution for an alternative, new way of thinking rigorously about biology that does not fall back on subjective modes of thinking. Er ... not so. Bizarrely, evolutionists lead the world at substituting teleology for objectivity. How so?

One major reason is the manner in which Natural Selection slipped seamlessly into the place of the Creator: the Natural Selector as the acceptable new face of the Great Designer. Darwin drew on examples of domestication as a major source of evidence for his theory. The Human Selector, full of purpose, was the model for the Natural Selector – just substitute the Natural World for the Pigeon Fancier. Predictably enough, anthropomorphizing Nature as your selector leads inevitably to the false supposition that there exists the quality of selectability, called 'Fitness' on the basis of which Nature selects. The pigeon fancier has specific features of the birds in mind when she makes her selection but Nature is mindless. 'Fitness' does not exist – it is another phantom construct of the human mind.

'Fitness' in the evolutionary sense of survivability cannot be a stable attribute of the gene, individual, or population because the likelihood that any of these 'selectable' entities will persist in time is also dependent on the nature of the surrounding circumstances, each of which (unlike a constant selector) varies more or less unpredictably in space and time. Features that increase the chances of survival in one set of circumstances may decrease it in others. The ability to bring more offspring to the reproductive stage, high 'reproductive fitness', almost always makes a positive contribution to survival simply because the statistical chance of total extinction is lower for a larger number. But numbers are no guarantee of survival – ask the Passenger Pigeon.

There is no selection, only differential survival, and since 'fitness' is defined as anything that promotes the chances of survival, both 'natural selection' and 'survival of the fittest' amount to no more than survival of the survivors, reflecting the uncreative emptiness of the continuous sieving of living things. The spontaneous generation of novelty *is* a real attribute of life, and the best that can be said for the lottery of elimination is that it frees up space and resources for more of it.

What follows on from 'fitness' is inevitably 'success', a concept given credence in ecology as well as evolution. While Professor Dawkins is chary of 'fitness', he has no problem with success. The problems come when we try to decide on how to measure it. Is it to be numbers? Total mass? Diversity? Well it all depends on your personal bias, because that's all success is. The concept is so subjective, we can't even agree on what constitutes success for people. Is it to be wealth? Happiness? Number of books sold? An organism is successful if you decide it is – 'success' as such doesn't have any real existence outside the mind. Coming back to evolution, the *only* reliable criterion of success is survival.

The evolution of an organism or a gene is its history, a long sequence of singularities, unique events like mutation, recombination, chance meetings and partings, matings, births and deaths, accidents of geography, and meteorology. Like history, evolution never repeats itself; like history, all attempts to derive a convincing set of underlying principles are doomed because every evolution is a one-off, unrecoverable, untestable, and unprovable.

This will never stop the pressure to simplify for human consumption the real myriad complexities of life forms and their achingly long histories by replacing the objective truth with metaphors more acceptable to the purpose-seeking mind. We may reject the flagrant teleology of aims, needs, intentions, and purposes, but exactly the same mental trick for rationalizing the world outside our heads re-surfaces in more muted forms as the crypto-teleology of a role, a task, a place in the plan – as 'function' (cue the endless series of dreary essays on 'structure and function', the needless rationalizing of real stuff from out there by providing it with reasons for its existence).

The explanations of the life mechanism produced by biochemists are powerful because they are about real components, assemblages of atoms whose existence in the here-and-now is solid and reliable, and this is deeply attractive to me as the only trustworthy reality. The moment we attribute 'function' to a familiar molecule we step off firm ground onto wobbly jelly, because function is a covert term for purpose and nothing else. Biological design is irresistibly seductive and unconsciously underlies all our thinking about the life mechanism. Constant vigilance, life long, is necessary in order to fight clear of the mindset that sees each molecule as having a job to do in sustaining life. Nobody gave them a job. They just exist, period. Well, can't we just accept the 'function' concept as an analogy that helps us make sense of a complex array of entities? Actually, no. This particular crutch is no longer needed and, I believe, actively hinders clear thinking. It no longer fits the facts. One: the same molecule may make a contribution to several processes. Two: the same process may be carried out by several different assemblages of molecules. 'Multi-functionality' and 'functional redundancy' are the norm, not the exception, and these terms are merely epicyclic attempts to patch up the 'function' concept.

The perceived weakness of biochemical explanation is that it relies on correlation. Explanations based on correlation are seen as lacking the strength of explanations based on causation. By contrast, the explanations generated by genetics are strong on causation – the genes are the onlie begetters, the ultimate prime causes, of the phenomena peculiar to living systems – but weak on mechanism. Geneticists generally paper over the glaring gap between the gene, with its exact chemistry, and the observable effect of which the gene is a contributory cause, using the concept of the Black Box. No coincidence, I suppose, that ignorance as emptiness is accounted black. Why not a White Box, a blank label on which the names of the components will eventually be written? Thankfully, causation is objective and can be understood independently of subjective notions like function (though few can resist corrupting it quite unnecessarily as 'gene function'), but the usefulness of explanations based on causal links generated by the genetic approach *is* limited by pleiotropy. If the White Box is small, containing only one or two components, say, chains of cause and effect are linear and the nature of the involvement of the gene in the life process in which its effects are detected is clear-cut. We all applaud the power of genetics. Mostly, though, the observable effect is a long way downstream of the gene, the boundaries of the White Box are undefined, and the chain of causation tortuous and indirect. It's not particularly useful to know one of the pleiotropic (i.e. indirect and secondary) effects of a specific gene.

Currently in my discipline, plant biology, there's a Gadarene rush to discover causal links between genes and the various life processes. It has recently become possible either to insert and express, or to knock out, the piece of genetic DNA of your choice, and those who work with plants are unconstrained by ethical concerns so long as their neophytes are confined to the lab. The general pattern emerging is that when you knock out a gene, or excessively over-express it, usually nothing happens. So, DNA sequences are no different from any other of life's chemicals: there is near universal multiple redundancy and, as shown by pleiotropy, any one gene can influence a number of processes, to varying extents.

My worry about the new biological explanations coming from genetics is that lacking a basis in the objective reality of chemical structure, and relying heavily on the fragmentary and indirect nature of any detectable downstream effects to suggest possible mechanisms, the choice of explanation is at the mercy of personal bias. What feels right is the intuitive explanation; intuition is prejudice acquired by long usage. All too often, explanations in biology that feel right intuitively do so because they reflect human values. They anthropomorphize, either imbuing the inanimate with covert animism – as in the case of 'function' – or infusing other forms of life with human qualities. More often than not the correct explanation in science turns out to be counter-intuitive, that is, not as we would have expected. That ought not to surprise us, but it always does. We ought to be much more distrustful of the plausible, the likely explanation.

One example is the way plants see – they detect light and respond appropriately. Every cell sees a range of colours, using different photoreceptors, but the most important is far-red, used for estimating how far away the nearest plant is. The same pigment, phytochrome, is also yoked to a biological clock in a system that measures the duration of the night with surprising accuracy – to the nearest 15 minutes – providing information on the time of year. This accuracy was discovered by experiments in which lights were switched on and off. How do plants decide exactly when night has started in natural daylight, which falls off gradually as the end of day is approached? This was a nagging puzzle for more than 40 years until Frank Salisbury, very near retirement, pointed out that the problem exists only in the minds of the investigators. The human eye adjusts to light intensity over an enormous range – a millionfold from midday to night vision – so we perceive the gradual onset of evening as spread over a very long time. The unaided plant pigment only detects differences over a tenfold range, so it goes from saturated to unable to see anything over a period of just 15 minutes at the end of the day (Salisbury 1981). By anthropomorphizing – thinking that a plant sees like a person – we made things seem more problematic than they are.

Recently, geneticists have found plant responses to light signals a fruitful area of enquiry. Sadly, their explanations of the results are riddled with prejudice and irredeemably warped.

When seeds germinate underground, the seedling in the dark shoots up rapidly, fuelled by stored food, burrowing out as a vegetable mole: hairless and pale with tiny leaves and a leading hook as a pusher. This is the 'etiolated' state and it is switched to the normal, large green-leaved, unhooked, light-fuelled state by red light. Geneticists collected mutants that couldn't respond to the light, and reasoned that the defect must be in a positive regulator of the light response. They also found mutants that stayed green in the dark and explained them as the result of loss of negative regulators of the light response.

Although this seems to be an essentially neutral model for explaining the genetic results, it isn't, of course. It's structured by intuition. The geneticist's vision is that in the plant in darkness the components linking the photoreceptor and the genes it controls are in an inactive, ground state. When there is a flash of light, the wiring between light receptor and genes is fired up and the genes get switched on, hence the response to light.

Others had already shown two important features of this system.

First, in the dark the plant is not in an inactivated state. Developmentally it is much more active – it's growing faster, for instance. Second, biochemically the components that connect the photoreceptor to the genes controlled by light are not in linear chains or anything like 'hard wiring'. They form a complex, tuned network of interaction, 'tuned' because every one is in a state of dynamic, furious turnover, constantly being created and destroyed, both in light *and* in the dark. Milling, seething turnover is the very spirit of all living systems. In this case it maintains cells in a state of high readiness for rapid change in response to light *or* dark.

A further point is that the etiolated state is a recent innovation in the long history of plants. Primitive plants stay green in the dark. The dark-adapted phenotype is a specialized add-on extra. If there is a 'ground state', it's the state in the light, not in the dark.

The current model is wrong because it is based on at least two implicit assumptions. First, that darkness is simply the absence of light, that is, no signal at all, and no signal means no response. Second, that transition from dark to light represents progress for the baby plant, moving successfully to the next stage of life, so the natural chronology is to activate the genes needed for its new life above the ground.

It sounds harmless enough, but in practice this model has blocked progress by distorting our understanding of what is going on.

Several of the 'negative regulators' of the light response turn out to be enzymes for the synthesis of a new class of growth hormones: brassinosteroids. They are only made in the dark, and are directly responsible for the faster growth in the dark. It's obvious that they are the 'dark hormone', but the literature insists they are inhibitors of the light response!

Worse, the investigation of whole sets of genes is foreclosed, thanks to the bias in the current explanation. There are more than 20 reviews on light-activated plant genes that are repressed in the dark. No one, repeat no one, is investigating the dark-activated, light-repressed genes whose expression generates the etiolated state. It can only be prejudice that ensures that darkness is unnoticed.

There are hopeful signs that we may be rescued from the perverting consequence of bias by a new type of biological methodology, the amassing of vast banks of data for its own sake, for example the genome projects. It may be mindless, but at least this ensures that it is value-neutral. In a recent study, more than 6000 random Arabidopsis (a weed) genes, one quarter of the genome, were screened for a change in activity in response to light. The authors were clearly startled to find that almost as many genes were upregulated in response to a dark signal as were induced by light (Ma *et al.* 2001). Sadly, but inevitably, they refer to them as 'light-downregulated genes'!

All our explanations of development are coloured by the prejudices of the prevailing culture, a culture which regards light, transition to later stages, and more of anything, as positive, but darkness, staying at the same stage, dormancy, and less of anything as negative.

Take, for example, the control of the onset of flowering. Intuitively, this is regarded as a positive step because at the beginning of their lives plants aren't able to produce flowers, so genes that act to delay the onset are grouped together as negative regulators of flowering. This description flies in the face of all we know about the nature of the control of flowering. For a large number of midsummer flowering plants, the 'long day' plants, the vegetative state is only, rather precariously, maintained if the plant is subjected to (delicious irony) a minimum period of darkness every 24 hours. Flowering is the default mode, and plants burst into bloom willy-nilly unless they receive their daily dose of darkness. It follows that in these plants negative regulators of flowering are much more likely to be positive regulators of the vegetative state (Hanke 1999).

Periodically, there are well-meaning attempts to justify the use of teleological reasoning as virtually harmless and an aid to thinking: 'You ask questions and you get answers you would never get were you not thinking this way' (Ruse 1989). Well, no thanks – those are not the sort of answers I am interested in. Teleology, the biologist's crutch, is bad not so much because it's lazy and wrong (which it is) but because it is a straitjacket for the mind, restricting truly creative scientific thinking. Attributing function tends to foreclose further consideration of the involvement of the gene, the protein, the membrane, the cell, or the organ, in other processes. It encourages us to split any entity into sealed compartments with different 'functions' instead of seeing the connections, instead of realizing it as an integrated whole. It's time to stop expecting that any part of the living world can be defined by function, however seductive or merely comfortable that feels. Feelings are deceiving. The truth, some of it counter-intuitive, is out there waiting.

References

Albertsson, P-Å. (2001). 'A quantitative model of the domain structure of the photosynthetic membrane'. *Trends in Plant Sciences*, 6, 349–53.

Dawkins, R. (1982). *The extended phenotype*. Oxford University Press.

Hanke, D. E. (1999). 'Accentuate the positive and eliminate the negative'. *Current Opinion in Plant Sciences*, 2, 423–5.

Ma, L., Li, J., Qu, L., *et al.* (2001). 'Light control of Arabidopsis development entails coordinated regulation of genome expression and cellular pathways'. *Plant Cell*, 13, 2589–607.

Ruse, M. (1989). 'Teleology in biology: is it a cause for concern?' *Trends in Ecology and Evolution*, 4, 51–3.

Salisbury, F. B. (1981). 'Twilight effect: initiating dark measurement in photoperiodism of Xanthium'. *Plant Physiology*, 67, 1230–8.

Colin McGinn

9

What is it not like to be a brain?

MATERIALISM is the thesis that facts about the mind are entirely reducible to facts about the brain. To be in pain, say, is to have one's C-fibres firing or for this brain state to realize a physically defined functional role. The usual objection to materialism, expressed in many different forms, is that it fails to do justice to the nature of the mind; it omits or distorts the distinctive character of mental phenomena. In this paper I shall not be pressing this objection to materialism. Instead, I shall invert the standard objection and argue that materialism fails to do justice to the nature of matter; it omits or distorts the distinctive character of physical phenomena. The symmetry of identity will play a crucial role in this argument.

* * *

Let me begin by listing a familiar set of characteristics commonly ascribed to mental phenomena, which are held to set the mind apart from the physical world. (By 'mental phenomena' I shall primarily mean conscious states and processes.) The list is eclectic and not uncontroversial; my intention is not to supply a full defence of it, but only to provide a foil for my own argument. The mind is held to be:

1. Unobservable – in the sense that mental states are not perceptible by means of the senses.

2. Asymmetrically accessible – in the sense that the owner of a mental state has a kind of immediate access to it that other people do not.

3. Subjective – in the sense that its nature is knowable only from a single 'point of view'.[1]

4. Non-spatial – in the sense that mental states do not take up a well-defined region of space.[2]

5. Subject-dependent – in the sense that mental states only exist for a subject of awareness.[3]

The usual claim, then, is that physical phenomena, such as brain states, do not exhibit these features, and hence cannot satisfactorily reduce mental states. Suppose the materialist maintains that pain is identical to C-fibre firing, so that there is nothing more to the state of being in pain than having one's C-fibres fire. The firing of C-fibres has the following characteristics: it is observable by means of the sense organs; it is accessible to oneself and others in the same way; it is objective in that it can be grasped from any point of view, not necessarily that of a pain-feeler; it is spatially defined; it could, in principle, exist without being experienced by a subject. The objection, accordingly, is that C-fibres are just the wrong kind of thing to identify with pain. If there were really nothing more to pain talk than C-fibres firing, then there would be no pain after all, since pain is defined by the opposite set of characteristics. The line of argument would be disputed by a materialist at various points, but prima facie it would seem that the anti-reductionist at least has a case that needs to be answered. The distinctive character of the mental certainly appears to be lost under such a reduction. It needs to be explained why it is that the reduction does not omit or distort the intrinsic nature of the mental. Let me summarize this objection by saying that materialism makes the subjective too objective. I shall take it that this is a familiar story, in one or another version.

* * *

The logical properties of identity statements have played a significant role in the defence and criticism of materialism. Thus the epistemic contingency and metaphysical necessity of identity statements such as 'pain = C-fibre firing' have figured heavily in these debates.[4] I want to draw attention to the least controversial property of the identity relation: its symmetry. This property allows us to say that if pain = C-fibre firing, then C-fibre firing = pain. And just as the truth of such an identity statement licenses us to say that there is nothing more to pain than C-fibre firing, so it licenses us to say that there is nothing more to C-fibre firing than pain. For C-fibre firing simply is pain, neither more nor less. If A = B, then there is nothing more to A than B and vice versa. C-fibre firing is not anything over and above pain. It has no properties not possessed by pain. It reduces to pain, collapses into it. It consists of pain. It has no reality beyond that of pain. C-fibre firing is constituted by pain. Pain is what C-fibre firing turns out to be. The essence of C-fibre firing is pain. But surely this sounds wrong: one wants to say that such an identification fails to do justice to the objectivity of C-fibre firing. If C-fibre firing were really nothing but pain, then it would not be observable, symmetrically accessible, conceivable from many points of view, spatial, and subject-independent. The identification makes the objective too subjective. It has the flavour of an elimination, not a reality-preserving reduction. If the reduction were correct, then C-fibre firing would not be an objective property of the world after all, contrary to appearances. The case is quite unlike, say, the identification of molecular motion with heat: here there is no strain in saying the molecular motion amounts to nothing but heat, since heat is not itself a subjective phenomenon. The objectivity of molecular motion is preserved in this reduction, whereas in the case of identifying C-fibre firing with pain we have an attempt to characterize an objective property in subjective terms. Thus the inversion permitted by the symmetry of identity results in an implausible reduction of the objective to the subjective – a loss of objectivity in the property we started out with.

It might be thought that this argument works only on condition that the reductionism takes the form of an identity claim. What if the reduction is formulated in terms of composition?[5] Suppose we say that water is composed of H_2O molecules; we cannot then symmetrically say that H_2O molecules are composed of water – such a claim permits a thesis of reduction of water to H_2O molecules. Thus we might analogously maintain that pain is (wholly) composed of the firings of C-fibres without committing ourselves to the converse claim. This certainly avoids the simple move from symmetry that I made against the identity formulation of reductionism, but I think that parallel problems beset the composition formulation too. First, if water is composed of H_2O molecules, then H_2O molecules are constituents or parts of water; equally C-fibre firings must be constituents or parts of pain if they compose pain. But how could something be literally a part of pain without being itself subjective? Consider a more familiar part–whole relation in respect of pain: I experience a complex pain resulting from banging my elbow against something hot. We might say that the resulting pain has both a collision component and a burn component; but these parts of my complex pain are clearly themselves subjective. If anything is a part of a subjective state, and not merely part of the neural correlate of that state, then it has to be subjective too. The claim is that pain is composed of nothing but C-fibre firings, but then C-fibre firings have to be the very elements that constitute pain, and hence must share its subjectivity. Just as a part of something objective must be objective, so a part of something subjective must be subjective. We would never allow that a putatively objective property might be wholly constituted by subjective elements, so why dispense a comparable leniency in the other direction?

Secondly, we can always derive an identity statement from a claim of composition. If X is composed of Fs, then there is some Y that is an aggregation of Fs such that Y = X. If water is composed of H_2O molecules, then there is an aggregation of those molecules such that the aggregation is identical to water. Water is not merely composed of H_2O molecules singly considered; it also is a collection of such molecules: it is identical to the aggregate of the elements that compose it. But then pain must be identical to the aggregate of the C-fibre firings that compose it, which is to say that the aggregate is identical to pain. This implies that (suitable) aggregates of C-fibre firings are as subjective as pains. When you put the C-fibre firings together you get something that is nothing over and above a pain. But surely an aggregate of objective elements should itself be an objective entity. Composition is not symmetrical, but it generates a symmetrical relation via the operation of aggregation. Again, the objective is collapsing into the subjective. So composition does not help the reductionist escape the argument. From now on, then, I shall persist with the simpler identity formulation.

*   *   *

A good way to get a feel for what I am arguing is to consider an imaginary school of philosophers who adopt a wholesale reduction of the objective to the subjective. These philosophers, call them 'mentalists', are troubled by the notion of objectivity; they find it hard to understand how there can be irreducible objective physical facts. Perhaps they think that the idea of such facts requires an 'absolute conception' that abstracts totally aware from their specific sensory point of view, and they cannot see how we could acquire such transcendent concepts.[6] Objectivity requires a 'view from nowhere', and they cannot conceive of such a detached view of the world. In any case, they find the idea of objective facts problematic for whatever reason. Yet they are not eliminativist, at least not officially: they agree that there really are objects that are square and electrically charged and made of neurones. They agree, too, that such facts can obtain whether or not they are being perceived by us (or by God); the mentalists are not idealists. What they insist is that every such fact is identical to a subjective fact. When an object is square, for example, they hold that this consists in the object's having a certain conscious state: there is something it is like to be square for the object. So-called objective properties are reducible to subjective properties, by way of suitable identity statements. They may not always know which subjective property a given objective property reduces to, but they are confident that there always is one. Every physical property is identical to a quale of some sort, known or unknown. This mentalistic metaphysics is different from panpsychism: panpsychism says that every object has some mental property, in addition to its physical properties; mentalism says that every physical property of every object is itself mental. There is no fact there is not something it is like to have.

What should we say about this startling mentalist doctrine? The obvious objection to it is that it is a reduction that signally fails to do justice to the objective physical world as we ordinarily understand it. If such a reduction were correct, then physical properties like being square or electrically charged would turn out to be unobservable, asymmetrically accessible, subjective, non-spatial and subject-dependent – given that mental states have these defining characteristics. Such properties would turn out to have all the proprietary features of the mental: they would be unobservable inner states of a subject of consciousness, not the publicly accessible objective properties we naively take them to be. And that is objectionably eliminativist, no matter what the official line of the mentalists may be. (One can imagine all the fancy footwork they would have to do in order to fend off the objection that we can see that being square is not an unobservable mental state just by looking at a square object. Compare the objection that we can see that a pain isn't C-fibre firing just by introspecting our pains.)

But if global mentalism of this kind is guilty of denying the objectivity of the physical, then surely local mentalism is too, albeit more narrowly. It is just as implausible to suggest that some physical properties are really mental as that all are. The global mentalists in effect treat physical terms as if they are natural kind terms for properties that will turn out to have a subjective essence – analogously to the way materialists take mental terms to be natural kind terms for properties that will turn out to have an objective essence. But local mentalists are really no better off: they take some physical terms – those for (some) brain states – to be natural kind terms for properties that will turn out to have a subjective essence. But this involves an implausibly subjectivist interpretation of an objective property, sacrificing all the objective features we naively ascribe to such properties. The subjectivist sin is not any less because it is localized. (Compare: some moral 'oughts' are reducible to naturalistic facts and some are not.) But local mentalism is the same doctrine as materialism, by the symmetry of identity, since the materialists precisely hold that some physical properties can be identified with subjective properties. They hold, for example, that C-fibre firing in the brain is identical to a sensation of pain felt by a conscious subject, and has not characteristics beyond those of pain. They hold that a physical property consists in a property defined by what it is like for its owner.

The trouble with materialism, ironically enough, is that it is not materialist enough about matter. It makes some pockets of matter too subjective in nature. It has essentially the same fault as global mentalism – a failure to respect the intrinsic objectivity of physical properties.

*   *   *

Let me incidentally note how odd the localized character of materialism is when seen for what it is. Offhand one might have thought that all physical properties are on a par, none having a fundamentally different essence from the others. But according to materialism some physical properties have a subjective essence while some do not. And it is not merely that physical states of the kidneys don't have a subjective essence while states of the brain do; some states of the brain itself have a subjective essence while some do not – despite the fact that all brain states consist of neurones and their firings. Not every brain state has a 'correlated' mental state. Whence this strange violation of the uniformity of nature? Induction would suggest that all neural states have a purely objective essence, since so many do; but when we come to a particular subset of them we allegedly find that they have a subjective essence. It is like discovering that some molecular motion is heat and some is not, despite the uniformity of the molecules and their motions. That would seem arbitrary and miraculous. But materialism finds itself countenancing something equally arbitrary and miraculous – the fact that some physical states but not others reduce to mental states. At the very least we need to be told what it is about this remarkable subset of physical states that makes them alone reducible to subjective states. Once we permit a robust notion of subjectivity this must seem a pressing question.

*   *   *

I want now to consider briefly some instructive parallels with the usual arguments surrounding materialist reduction.

What is it like?

Thomas Nagel argued that there is something it is like to undergo conscious experience, and that this something is accessible only to those beings who enjoy similar experiences.[7] He then argued that brain states are not defined in such terms; they are accessible from a variety of experiential standpoints. Hence the claim of reduction is flawed: we cannot find a place for the subjective in our objective conception of the world, including the brain. Inverting this, I insist that it is part of the very definition of a physical state of the brain that it is objective, in the sense that it is knowable by beings with the right intelligence irrespective of the particular types of experience they enjoy, so that we cannot reduce such states to states bearing the marks of subjectivity. Just as it is important that there is something it is like to be a bat, so it is important that there is nothing it is like to have a bat's brain; that is, it is important that bat brain states are objective properties of the world. We do not want to collapse bat brain states into purely inner processes. Indeed, if we did we would not know what properties the bat's brain has, since we do not know what kinds of experience it has – which is absurd. If facts about bat brains are identical to facts about what it is like to be a bat, then such facts are not knowable without sharing the subjectivity of a bat – that is, they are subjective facts. But they are not subjective facts, since the neurophysiology of bats is knowable by beings other than bats. Thus the objectivity of a bat's brain is just as inconsistent with materialist reduction as the subjectivity of its experience is. Either we deny the subjectivity of the experience in order to sustain the reduction to brain states, or we deny the objectivity of the brain in order to conform to the subjectivity of experience. If Nagel is right about the inherent subjectivity of experience, as I think he is, then materialism results in a denial of the objectivity of matter. The B-fibres that are identified with the bat's subjective experience will turn out to have all the subjectivity of those experiences, and nothing more.

The knowledge argument

Frank Jackson's Mary is said to know about all the physical facts without being thereby apprised of all the facts about the mind; she cannot deduce subjective facts from her comprehensive knowledge of objective facts.[8] When Mary emerges from her black-and-white room armed with complete knowledge of neurophysiology she learns something new when she first enjoys an experience of red. So what she learns was not prefigured in what she already knew. Hence materialism is not a complete theory of the mind. Now consider Maisie: she knows all the phenomenal facts – all the facts about her own mental states and their interrelations. She spends the early years of her life floating in a vat enjoying her own phenomenology, thinking about her experiences, classifying them, revelling in them. There is little outside distraction from her inner world; specifically she is taught no physics, including neurophysiology. One day she is removed from her phenomenological vat and forced to learn physics. In the course of her studies, at which she proves remarkably adept, she learns all about her brain, including the correlates of the phenomenal states she knows so well. Does she thereby learn anything new? She used only to know what experiences of red were like; now she also knows all about the R-fibres that underlie these experiences. Well, it certainly seems like she learns a new fact – that she has R-fibres that correlate with her familiar old experience of red. Moreover, she learns a fact of a new kind – an objective fact, as distinct from a subjective phenomenal fact. Therefore R-fibres cannot be identical with experiences of red. Just as subjective facts cannot be deduced from objective facts, this creating a knowledge gap, so objective facts cannot be deduced from subjective facts, this also creating a knowledge gap. The gap is as large whichever direction you approach it from. Maisie is as ignorant as Mary before both their life-styles change. (Of course, there are replies to the knowledge argument, which I will not go into here,[9] and replies to these replies, but my point is just that the argument cuts both ways.)

What God had to do

Saul Krepke maintains that when God created pain he had to do more than create C-fibre firing – whereas to create heat is sufficed to create molecular motion.[10] At least that is our strong intuition. If the intuition is correct, then the mind is not necessarily supervenient on the brain. Pain must be something over and above C-fibre firing. Accordingly, zombies are conceivable: beings in some possible world who share our physical properties but differ from us in having no mental states at all.[11] There is a modal and ontological gap between C-fibres and pain, marked by the tasks God had to perform to make a world like ours. But it is no less intuitive to make the opposite point; in order to create C-fibres it was not enough for God to create pain – he had some additional work to do in order to bring C-fibres into the world. Accordingly, mental states do not logically determine physical states: there is a lack of supervenience here, and disembodiment seems logically conceivable. That is, after having created pain it was up to God (i) what physical state to correlate with it, and (ii) whether to correlate any physical state with it. There is thus a modal and ontological gap between pain and C-fibres, marked by the extra effort involved in producing the latter after producing the former. Their arguments are exactly parallel.

Now it is not that I myself subscribe to either of these arguments: I actually believe that the connexions here are necessary despite our modal intuitions to the contrary.[12] What I am saying is that the arguments are exactly parallel, so that anyone who accepts them one way round has to ask whether to accept them the other way round. In particular, those who believe in the possibility of zombies need to ask whether they also believe in the possibility of disembodied minds.

The epistemic interpretation

A standard reply to the above anti-reductionist arguments is that they confuse ontology with epistemology.[13] Our concepts of pain and C-fibre firing may indeed be distinct, but it does not follow that they denote distinct properties; and mental properties may be reducible to physical properties without the concepts that denote them being reducible to physical concepts. The idea is that 'subjective' and 'objective' are predicates that apply to properties or facts only under certain descriptions or concepts. A property P might be subjective under the description 'pain' and objective under the description 'C-fibre firing'. The ontological gap that seems to separate the subjective from the objective is really just an epistemic gap between the concept of pain and the concept of C-fibre firing, not a gap between the properties themselves. Now a lot can be said about this form of reply, but I want to make only one point relevant to the argument of this paper.[14] The epistemic reply to the claim that materialist reduction fails to do justice to the nature of the subjective is that pain is only subjective as so described; it is subjective de dicto but not de re. When we redescribe pain as C-fibre firing we can see that it is really an objective property in itself (de re). The characteristics I listed at the beginning as distinctive of the mind belong to it only under a mental description – they apply to pain only de dicto. In effect, they all generate opaque contexts. But if this is true of the subjective it must also be true of the objective: properties are only objective under certain descriptions, and never de re. When I redescribe C-fibre firing as pain it ceases to be objective, save relative to that physical description. C-fibre firing is only objective under a description and not in itself. And similarly for any other apparently objective property: we cannot say that having an electrical charge is objective de re but only that it is objective de dicto, since the notion of objectivity is being interpreted merely epistemically. If I think of an electrical charge under the description 'that which causes pain in humans', then I consign it to the class of subjective facts. According to this view, it makes no sense to attribute objectivity (or subjectivity) to states of affairs in themselves. We cannot even say that a universe in which there are no minds contains purely objective facts, unless reference is made to our current concepts. This seems very odd. Surely it is an inherent intrinsic de re fact that physical states in general are objective states, in the sense I spelled out at the beginning. It is not that they become objective only when we decide to describe them in a certain way. It is absurd to suggest, say, that physical objects are only spatial under a description and not in themselves. It is in the very nature of physical facts that they are objective.

The point I am making is that materialists implicitly adopt an invidious attitude towards the subjective and the objective: they are only too happy to assert that properties are subjective only under a certain description, but the parallel move for objectivity looks distinctly unappealing once its implications are appreciated. Yet it is this move that is necessary if we are going to object to my arguments by saying that C-fibres can be objective under that description but not under the description 'pain'. That is no way to protect the robust objectivity of physical facts. The plain truth is that if pain is allowed to be robustly subjective, in the de re sense, then identifying C-fibre firing with pain results in divesting this physical process of its vaunted objectivity. This objectivity cannot be plausibly restored by retreating to the thesis that C-fibre firing is objective only de dicto – on pain of making all objective facts similarly weakly objective. So the epistemic interpretation fails to deliver a robust notion of objectivity, just as it fails to provide a robust sense of subjectivity. The latter is tolerable to a materialist, given his ontological biases; but the former is surely highly unpalatable to the materialist.

There are three options here: (i) the characteristic marks of subjectivity or objectivity are not possessed at all, so that we end up with eliminativism either about the subjective or the objective; (ii) these characteristics are possessed only in the de dicto sense, so that we end up denying that anything can be subjective or objective in itself; (iii) we allow that properties can be subjective or objective de re, so that we end up either distorting the nature of the subjective or (the plaint of this paper) distorting the nature of objectivity. Assuming that we want to avoid eliminativism, we have the result that you cannot be an identity theorist who respects the robustness of subjectivity and respects the robustness of objectivity. If you identify a physical property with a genuinely mental property, then you cannot avoid an unacceptable subjectivization of the objective. The trouble with materialism is that it does not take the objectivity of matter seriously enough, despite its overt intentions.

* * *

What should we conclude from this discussion? I think we can conclude that classic type-identity materialism is false, either the central state version or physicalistic functionalism (and behaviourism). But I do not think we can conclude either that mental states are irreducible or that there are merely contingent connexions between mental states and brain states. Maybe mental states are reducible to something that does not have the marks of full-blown objectivity, unlike C-fibre firing and its kin.

Maybe new properties could be discovered that both reduce mental states and are not themselves objective; and even if they could not be discovered they might nevertheless exist.[15] And there may be necessary connexions between pain and C-fibre firing even though it is not possible to identify the two; these may be distinct properties that are non-contingently connected. The nature of the necessary links might be hidden and not inferable from our current concepts, but they might exist anyway.[16] So nothing I have said entails a rejection of supervenience or an acceptance of irreducibly mental properties. The correct view of the mind–body relation is left open by what I have argued. All I have contended is that the usual kinds of materialistic identity theory are committed to an unacceptably subjective conception of the physical world. As Kripke remarked after presenting his own modal argument against the identity theory, the mind–body problem is 'wide open and extremely confusing'.[17]

Notes

1. See T. Nagel, 'What is it like to be a bat?', reprinted in his *Moral questions* (Cambridge University Press, 1979), and *The view from nowhere* (Oxford University Press, 1986).

2. I discuss the non-spatiality of the mind in 'Consciousness and Space', in *Conscious experience* (ed. T. Metzinger), pp. 149–63 (Imprint Academic, Shoningh, 1995).

3. J. Searle expresses this point by saying that conscious states have 'first-person ontology': see his *Rediscovery of the mind* (MIT Press, Cambridge, MA, 1992).

4. On the epistemic contingency of psychophysical identity statements, see J. J. C. Smart, 'Sensations and Brain Processes', *Philosophy Review*, LXVII (1959), 141–56. On the metaphysical necessity of identity statements see S. Kripke, *Naming and necessity* (Harvard University Press, Cambridge, MA, 1980).

5. Thomas Nagel suggested that I consider this line of defence.

6. On the availability of the 'absolute conception', see B. Williams, *Descartes: the project of pure enquiry* (Penguin, Harmondsworth, 1978); Nagel, *The view from nowhere*; and my *The subjective view* (Oxford University Press, 1982).

7. Nagel, 'What is it like to be a bat?', and *The view from nowhere*.

8. F. Jackson, 'Epiphenomenalism and Qualia', *Philosophical Quarterly* (1982), XXXII, 127–36, and 'What Mary didn't know', *Journal of Philosophy* (1986), LXXXIII, 291–5.

9. I make a few remarks about one standard reply below. For a much fuller discussion see my 'How not to solve the mind–body problem', in *Physicalism and its discontents* (ed. B. Loewer), pp. 284–306 (Cambridge University Press, 2001).

10. Kripke, *Naming and necessity*.

11. See D. Chalmer, *The conscious mind* (Oxford University Press, 1996), in which the alleged conceivability of zombies plays a pivotal role.

12. On the idea of hidden necessary connexions, see my *The problem of consciousness* (Basil Blackwell, Oxford, 1991), esp. pp. 19–21, and my review of Chalmer in my *Minds and bodies* (Oxford University Press, New York, 1997).

13. For example, P. Churchland, 'Knowing qualia: a reply to Jackson', reprinted in P. M. Churchland and P. S. Churchland, *On the contrary* (MIT Press, Cambridge, MA, 1998). For a different version of this line see B. Loar, 'Phenomenal states', in *The nature of consciousness* (ed. N. Block, O. Flanagan, and G. Guzeidere), pp. 597–616 (MIT Press, Cambridge, MA, 1998).

14. The obvious question is what makes concepts different if not the properties they express: see my 'How not to solve the mind–body problem' for more on this.

15. I believe, for reasons not entered into in this paper, that the most likely hypothesis is that the properties that are needed to solve the mind–body problem are in principle inaccessible to the human mind; my present point is just that a reduction might exist that we shall never discover. See my *The problem of consciousness* for a discussion of the insolubility thesis.

16. See note 12 for references on this.

17. Kripke, *Naming and necessity*, p. 155, note 77.

Tian Yu Cao

10

Ontology and scientific explanation

OVER millennia, we humans have invented and developed a multiplicity of ways to understand whatever in the world intrigues us. These many modes of understanding include mythical projections, religious teachings, metaphysical speculations, magical association, hermeneutic readings, mathematical and logical reasoning, and scientific explanation. What differentiates scientific explanation from other modes of understanding is a question standing at the very centre of the philosophy of science, a question to which philosophers of various persuasions have different answers. In this short piece I am not, however, going to examine these diverse opinions beyond offering a few brief comments. Rather, my major concern is to give an outline of my own view of scientific explanation, which I will then illustrate by several examples taken mainly from physics.

Sources of explanatory power

Scientific explanation differs from mere description in that it yields new understanding of the explanandum, that which is to be explained in terms of something else, the explanans. Scientific explanation differs also from other modes of understanding in the complexity or abstractness of the explanans, which itself, in turn, forms a complex or abstract explanandum. An explanans/explanandum may be a particular phenomenon, may be a regularity or law, or may be a theory about such phenomena or regularities. The explanans/explanandum may also be the most general perceived or imputed features of the world, such as its having both unity and diversity, or even the genesis and evolution of the universe as a whole.

Philosophers generally agree that to qualify as scientific, an explanation must be empirically adequate, must be logically consistent, must not be designed for the particular case (non-ad hoc-ness), and so forth. But our most important question – the question on which philosophers divide – is why and how the explanans carries explanatory power over the explanandum. Put another way, what is the source of the explanatory power projected by the explanans onto the explanandum such that all criteria of being scientific are met.

The common accounts about the sources of explanatory power are three. First is the neo-Platonist account favoured by many theoretical physicists, in which mathematical entities, their representations, and the patterns of the logical structures derived from their definitions – in particular gauge symmetries and D-branes – are taken as the true reality or as manifesting the hidden essence that exists beneath overt phenomena. By revealing themselves to be the real nature of various phenomena, these mathematical entities constitute the universal foundation of physical theories, from which the complexity of the universe can be explained in a consistent way (Radicati 1984).

Second, it is quite popular among scientists and philosophers to locate the sources of explanatory power in causal structures, that is, structures produced by causal interactions and propagated through causal processes governed by causal laws. Here 'cause' may be understood either in the sense of antecedent happenings or in the sense of properties of constitutive entities, such as explaining the hardness of a diamond through its constitutive carbon atoms fitting together neatly. In this latter case, the causal explanation is usually taken to be the paradigmatic case of reductive explanation (Salmon 1984; Weinberg 1995).

Third, according to the deductive–nomological (D/N) model of explanation, the sources of explanatory power lies in the deducibility of the explanandum, under certain conditions, from some law-like regularities under which the explanandum is subsumed. Thus the more general the regularities are, the richer the resources of explanatory power they possess (Hempel 1965; Watkins 1984).

The difficulties with these accounts are many. Consider the neo-Platonist account: why should we believe that the reality and essence of the physical world actually lie in purely mathematical entities, when from those mathematical constructions alone we get nowhere without physical inputs.

Next, the causal account could be acceptable only if it offered a proper account about the source of causal power, but it does not. As to the D/N model, it, too, gives no account of what the connection is that makes it possible to deduce the explanandum from the explanans. Moreover, none of these accounts provide a proper understanding of scientific explanation, an understanding that would give it proper extension and make its distinctive features understandable.

In my view, the source of explanatory power lies in ontology, which means the primary entities that are assumed to exist in the domain under investigation. The primary entities are those from which all appearances (other entities, events, processes, and regularities) are derivable as consequences of their properties and behaviour; these primary entities display regularities and obey laws, the so-called fundamental laws in the domain concerned. Since, by definition, these primary entities are the site and embodiment of the essence and reality of all phenomena, possessing both the causal power of the explanans and the deducibility of the explanandum, this account has whatever merits the other accounts have. But let me elaborate a bit on what scientific explanation is, and why it follows that ontology, and only ontology, provides explanatory power.

If the purpose of scientific explanation is to understand the meaning of the explanandum, we have to clarify just what the vehicle is through which the meaning of whatever it is that is to be understood can be understood. On this crucial question I follow Mary Hesse (1965; see also 1963, and Arbib and Hesse 1986): ultimately we have to rely on the vehicle of metaphor. That is, if by means of a chain of metaphors, something can be connected with the reality of everyday life through a kind of structural similarity possessed by each link of the chain, then its meaning is understandable. On this view, then, scientific explanation can be defined as metaphorical redescription of the explanandum in terms of an ontology, whose properties (as the source of causal power) and behaviours (that display regularities and obey laws) are assumed to be understandable through a chain of historically developed metaphors in science.

Thus, on this view the fundamental task in scientific explanation is to find the primary entities (and their properties and behaviours and the laws and principles that govern them) in the domain under investigation, and to take this ontology as the metaphor by which to redescribe, in its terms, whatever is to be explained. This entails that we employ several – even many – different prototypes of explanation because a different domain demands a different ontology for its metaphorical redescription. But how are we to proceed when the explanatory task involves two domains at different levels in the hypothetical hierarchical structure of the world? And, what if the task is to explain the genesis of the universe as a whole? To this last question I come at the end of this essay; to the former questions I turn now.

Types of explanation

It is a truism that science, among other merits, satisfies our curiosity by seeking to explain many matters in the world as we experience it, whether through our unaided senses or with refined instruments.

These range from astronomical phenomena to the subatomic world, from the genesis of the universe to the emergence of life and consciousness. Less commonly appreciated is that science also provides differing prototypes for explaining. For example, mechanics, according to which motion of a material body is the result of some force(s) acting upon it, offers a prototype for causal explanation. Indeed, many take causal explanation as the only legitimate way of explaining. But when we move to other domains, such as living organisms or quantum fields, where the primary entities are not classical mechanics' inert material bodies, the basic metaphors offered by the ontologies in these other domains provide different ways of explaining. Thus, for example, different branches of biology have offered the prototypes of functional explanation on the one hand, and evolutionary explanation on the other, both of which are distinctively different from causal explanation, at least if the latter's meaning is not overstretched beyond physical mechanisms. Another prototype of explanation emerged from quantum physics. It has many features originating from the quantum ontology that are absent in, or even in contradiction with, traditional causal explanation. Whether we can put this quantum prototype under a more general – but yet not over-stretched – concept of causal explanation is a subject for debate, to which I will return.

The notion of 'prototype' is relevant to explanation because the way of explaining suggested by the basic metaphor offered by the ontology of one branch of science may be applicable to another domain in that science, or in an entirely different science, if the explanandum in that other domain shares some features with the original metaphor. Thus functional explanation may be applied to social phenomena, such as productive forces and relations of production, economic structures, and legal-political institutes, etc. Similarly, the evolutionary metaphor can be used to explain special features of some physical systems, as, for example, the applicability of the anthropic principle in cosmology. This principle, so unreasonable or mystical/theological from the perspective of the 'evolution' of a unique universe, is potentiated when our older picture of the universe as a unique system is replaced by a new evolutionary picture of a multiverse.

What, then, about the radically new type of explanation suggested by quantum physics? It is adopted not only by physicists working in contemporary physics, but also by some philosophers, who use it to support the D/N model of explanation. For this reason, we first have to clarify the notion of causality and the notion of law-like regularities, and then their relationship, before moving to a discussion of this new type of explanation per se and its relationship with the D/N model.

The notion of causality has many qualifications. Causality may be singular or general, necessary or sufficient, deterministic or statistical. It may be equated to functional dependence expressed by general correlations or based on causally effective properties. Let us start with positive singular causation: the occurrence of a singular effect is the result of the occurrence of a singular cause. Obviously, such a case can be expressed as a special case of correlation between the cause and the effect. Suppose, however, we take this positive singular case to be special, not merely in the statistical and perfect correlation senses, but also special in the sense that what underlies positive singular causation are some properties actually possessed by the entities involved in the cause and effect. Then causality can be understood as a special kind of correlation, which is to be distinguished from more general correlations where no causally effective properties can be detected or assumed to exist as the foundation. If causes come in kinds which enable correlations to fix the kinds of effects they have, then we have moved from singular to general causality, and causal correlations become laws of nature. In laws of nature, causally effective properties possessed by the entities involved in general causality feature prominently and, in fact, make the laws of nature possible (Mellor 1995).

In this way of looking at the relationship between causality and laws of nature, laws of nature have their roots in causality, and thus are different from other observed generalizations or regularities. Then the success of the D/N model in explaining phenomena by appealing to laws of nature or law-like generalizations can itself be understood in causal terms. Moreover it helps us to understand why laws must be put in the context of a theory to be understandable: entities and causally effective properties that make these entities instantiate the laws are specified by a theory. In other words, only a theory can specify an ontology which, as the carrier of causal structure, functions as a conceptual foundation for understanding the laws.

Standing in diametric opposition to the foregoing account of the source of explanatory power in science is the all-too-familiar positivist view that laws of nature are nothing more or less than universally valid correlations. It takes these bare correlations as the ultimate basis for our integration of experience. In this view, causality, too, can be nothing more or less than functional dependence of explanandum on explanans. Thus 'causal' means simply and solely capable of being expressed or defined in terms of laws or law-like regularities. Similarly, on this view properties are not taken to be causally effective. Rather, properties are themselves defined by laws, indeed owe their existence to laws: only those properties that occur in laws of nature can be taken as actually existing. Clearly, the D/N model follows directly from this view, as does also the precept that the discovery of laws of nature and general principles is the most important task in physics, and that the specification of ontology is not merely incorrigibly speculative but entirely dispensable.

This view, and the D/N model of explanation along with it, have been greatly reinforced by quantum physics, where (i) prevailing randomness defies singular causality; (ii) the impossibility of any space-time picture of the microscopic world further undermines the traditional conception of causality, according to which causes and effects are related in space and time by dense sequences of intermediate causes and effects; and (iii) all that is epistemically accessible to us are regularities expressed in terms of mathematical correlations, through which probabilities of observable quantities are calculable in many cases with amazing precision, and thus whether ontology may be posited at the outset is questionable. Since causality, on this view, involves nothing more than functional dependence, it is not self-contradictory – although perhaps also not particularly meaningful – to speak of explanation in quantum physics based on the notion of 'statistical causality'. For this reason, causal or not, this type of explanation is called by many simply statistical explanation. Clearly, statistical explanation is different from traditional modes, which aim at explaining single events.

Now an interesting question concerns whether we can define the notion of statistical causality in a different way, in a way that is based on my view of the relationship between causality and laws or regularities, such that the explanatory novelties brought by quantum physics can be ontologically derived from the novelties of the primary entities in the quantum domain – it being understood that, epistemically, the probabilities of observable quantities are our only effective means for access to the novel quantum ontology.

How to explain quantum phenomena and space-time

One must acknowledge right off that there are good reasons for rejecting the possibility of giving quantum phenomena a causal explanation, where causality is taken, as I take it, to be based on factual properties, properties possessed by existing physical entities. First, no property can be consistently ascribed to a constituent entity of a quantum system before that property (parameter value) has been measured. This is clearly demonstrated by experimental tests of Bell's theorem, a very general relation between quantum-mechanical variables derived by J. S. Bell in 1964 and 1966 (see Bell 1987). Second, because of the act of quantum jumps, no such entity can even be conceived as having a continuous existence. The lack of continuous existence of any quantum entity refers not only to the spatial and temporal aspects of its existence, but also to its existence per se. Third, the very existence of quantum jumps also challenges the causal efficacy of physical properties, because no physical process can be visualized through which the power of the cause can reach from one end to the other end of the causal chain. Fourth, and most ominous, the inherent randomness of quantum jumps destroys any possibility of singular causality, which in my view is the starting point for defining causality.

All these reasons for renouncing causal explanation in the quantum domain are granted. Also granted is the fact that the only means through which we can have access to quantum reality is abstract mathematical formalism expressing general laws and principles, which cannot be taken as representing physical processes visualizable in spatial-temporal terms. Nonetheless, I argue that the type of explanation suggested by quantum physics remains based on the notion of extant primary entities possessing causally effective properties.

First, quantum physics cannot by any means dispense with such properties as charge, mass, and spin, which figure prominently in laws governing the behaviour of primary entities in the quantum domain.

Second, there exist mechanisms and processes through which interactions can be realized and influences can be transmitted. Although these mechanisms and processes are not visualizable, they, nevertheless, can be defined mathematically in a very precise way.

Third, all physical processes in quantum physics are governed by laws of nature, and all entities involved in these laws come in kinds.

Fourth, as a result of mathematical definiteness, the probabilities for various occurrences can be calculated in a rigorous and precise way. Thus the function of a cause is more than statistically relevant in the limited sense of being able to raise the chance of the potential effect. Rather, the causal link between single cause and single effect is definite and calculable. The link thus defined is statistical, thus the appropriateness of the name 'statistical causality', but now in a deeper and more meaningful sense than the previous sense of rejecting singular causality. The crucial point here is that, what in the quantum domain function as singular causes and singular effects are, ultimately, the primary entities characterized by specific physical properties, and these entities have only statistical existence themselves.

Does this position concede too much to the 'positivists'? There are, after all, some speculative physicists and many philosophers who still try hard to achieve a reduction of the quantum ontology into a classical one at a deeper level.

Is it legitimate to take the quantum ontology with only statistical existence as a brute fact from which to conceptualize the quantum situation and to ground the special types of explanation in the quantum domain? A point I wish to stress here is that even if we succeed in reducing the quantum ontology to a classical one of definite existence at a deeper level, it will not change the nature of quantum ontology at the quantum level. Moreover, no explanation can proceed without any brute fact, for otherwise we are caught in an infinite regress.

With the notion of statistical causality, defined now in line with my own view of the relation between causality and laws of nature, the explanation as derivability, as it appears in the quantum domain, can be easily recast in causal terms, and all its novelties can be traced to the novelties of the quantum ontology.

Another interesting application of the ontological approach to explanation that I suggest above involves the explanation of space and time, or space-time. When I mentioned the collapse of any space-time picture in quantum physics, I did not mean that the quantum world existed outside of space-time. What I meant was only that its space-time behaviour differed from the classical world: discontinuity replaced continuity. Whatever happens in the quantum world still happens in space-time. In fact, the theoretical structure of quantum physics, and quantum field theory in particular, is largely dictated by the Minkowskian metric, which specifies the space-time behaviour in the absence of gravity, without which no commutation (or anticommutation) relations and microcausality can be defined. If whatever exists (or happens) exists (or happens) in space-time, then, space and time are primary notions not explainable by others. This approximates Kant's thesis that no experience or knowledge of the physical world would be possible without space and time. But what Kant had touched upon was only epistemological necessity, not ontological priority. Kant's argument has not really ruled out the possibility of giving space-time an ontological explanation. In Kant's view, space and time are necessary forms for having access to the phenomenal world. According to Mach and Einstein, however, and in particular in the spirit of general relativity, these relational entities can have an ontological explanation in terms of interacting bodies, or metric tensor fields, or ultimately the gravitational field (affine connections).

Briefly, the reasoning is as follows. In general relativity, since the field equations are generally covariant, the points of the four-dimensional space-time manifold have no identity, but function only as place-holders of a relational structure connecting them together. Here the relational structure refers to the chrono-geometrical structure described by metric tensor fields, solutions to the field equations.

If the spatio-temporal relations are not specified by a reference to space-time manifold points, but rather to metric tensor fields, then manifold points in themselves have no spatio-temporal meaning. They can have such a meaning only when a metric tensor field is assigned to them. Thus it is not true that a manifold endowed with some given differential and topological structure represents space-time, or space-time can exist on its own footing. Rather, space-time as a relational entity is constituted by metric tensor fields. This means that the metric tensor field is solely responsible for defining space-time relations.

In contrast to the Newtonian space and time, which can be taken to have an independent existence without physical interactions with other entities, and also in contrast to the Leibnizian spatial and temporal relations, which are not physical entities (existing independently) at all, the metric tensor field is a dynamic entity interacting with all physical entities, including itself. It is this universal coupling, which determines the relative motion of the objects we want to use as rods and clocks, that allows us to give it a metrical interpretation, which in turn justifies its defining role in constituting spatio-temporal relations.

Frequently people conflate the metric tensor field with space-time or its chrono-geometry. Typical examples can be found in talking about 'the excitations of space-time itself'; or 'the quantization of geometry', which are quite fashionable among quantum gravity experts. This, however, is not appropriate, because such expressions render invisible the important difference between the constituting agent (i.e., the ontology endowed with causal power for explanation) and the constituted relational entity. The difference is important because its recognition helps clarify which theoretical entity should be taken as physically primary, underlying, and explanatory in the discourse of gravity and space-time. And clarity on that point is, in turn, important for choosing what to be subject to quantum theoretical treatment in quantizing gravity.

How to understand the unity and diversity of the world

Thus far I have presented the ontological approach to explanation with an emphasis on the primary entities within a domain under investigation. Is this approach also able to address issues involving two or more domains at different levels in the hierarchy of the world? This is necessary if we are to understand, ultimately, the unity and diversity of the world.

These questions are not new, and have been addressed over centuries by reductionists. Of course, no wise reductionist would deny that in the course of cosmic evolution, new entities and new phenomena have frequently emerged. But they argue that the true unity of the world can be shown only by an account on reductionist principles. First, the emergent entities and phenomena are always causally connected with the entities from which they have emerged. Second, and this is more important, the laws governing the behaviour of the emergent entities are explainable by the more fundamental laws that govern the behaviour of the underlying entities. For example, Francis Crick argues that our entire mental life is caused by the behaviour of neurones, and that 'behaviour' is simply the rate at which the neurone 'fires' electrical pulses into other neurones. And that rate, the reductionist argues, is dictated and thus explained by physical laws. In fact, the reductionist claims, there are no unbridgeable gaps between different entities, or between different properties and aspects of entities, that cannot be closed by causal connections.

Among contemporary physicists the best known and most forceful exponent of reductionism is Steven Weinberg (1995). For him, the old dream of physicalism remains a firm conviction: science is the search for the simple and universal laws of nature that explain why the world is the way it is. These ultimate laws and principles lie at the starting point of all chains of explanation and thus serve as the common source of all explanation. In this the science of physics has a privileged place: physics reduces the world of physical phenomena to a finite set of equations and principles, to which all laws and principles of other sciences can in turn be reduced. Here the conviction of the unity of nature becomes an argument for the unity of science under the hegemony of physics.

In this spirit, Weinberg acknowledges that there is emergence of life and mind. Yet, he insists that the rules that mental phenomena obey are not independent truths, but must follow principles at deeper levels. Apart from historical accidents, he stresses, the nervous system has evolved to what it is entirely because of physics. But physicalism is not the only contemporary form of reductionism; 'informationism' – the gospel that at bottom the world is not material but consists of dis-embodied information – has gained a remarkable ascendancy in recent decades. This form of reductionism appears in attempts to eliminate the mind in terms of artificial intelligence or by identifying the mind with brain programs. There are of course also middle-level reduction-ist accounts that take neuronal processes as the causal basis to give a causal emergentist explanation of the mind, in terms of, for example, the binding of neurone firings, or re-entrant excitations in neurone networks.

The reductionist version of monism is, understandably, rejected by dualists, who take the mental as irreducible to the physical and thus as enjoying an explanatory priority. Hilary Putnam (1993, 1994), for example, argues that perception is a transition from potential to actual awareness, and this is radically different from the transition of a phys-ical system from one state to another state. For Putnam, the semantic properties of thought as they involve meaning and truth cannot be reduced to, or even accounted for, in terms of cause–effect. These semantic properties are intrinsically intentional and shaped in large part by social usage in a speech community. Thus mind or reason cannot be naturalized in the sense of being exhausted by the natural sciences.

Putnam's dualism is based on his arguments against a reductionist view of explanation. For him, even if the behaviour of a system at one level is deducible from the descriptions of something at a lower level, this deduction is by no means an explanation: the relevant features in this way of deduction would always be buried in a mass of irrelevant information and cannot be brought out. In a head-on confrontation with Weinberg, Putnam argued that the laws of a higher level cannot be deduced from the laws of a lower level. The higher level laws of emergent entities have integrated into themselves some initial and boundary conditions that are essential to the emergence of new enti-ties with novel features. Such boundary conditions are indispensable to the description of the higher level – indeed to the constitution of the emergent entities – but are accidental from the point of view of the lower level. For this reason, the laws of a higher level are in fact auto-nomous vis-à-vis the laws of the lower level. All upper-level phenom-ena and laws have a basis in physics. Yet they cannot be deduced from the laws of physics. They must be compatible with the laws of physics. But that mere compatibility means that physics has receded to the background and become irrelevant in explaining particular phenomena or laws in other domains.

Based on these anti-reductive considerations, Putnam argues that the intentional mind is not reducible to representational systems treated in computer and brain sciences. Thus Putnam disqualifies Crick's neurone firing rate representation, Edelman's neural assembly architecture representation, and Chomsky's inner template as hard-wired in the brain by evolution for natural language, semantics, and other conceptual aspects of our mental representations. Putnam allows that this kind of reductive pursuit may be of some limited utility, but can provide no real understanding of the intentional. The fatal defect of representationism, according to Putnam, lies in its dealing with the link between representations and the external world only causally, but not cognitively, thus having missed the essential features of the inten-tional.

It is Putnam, rather, who, like so many other freedom-loving philosophers, is led by his anti-determinism into an unwarranted anti-causalism. But although Putnam does not realize that the inapplica-bility of causal explanation in the domain of intentionality is due to the fact that the primary entities in this domain have features different from those of mechanical systems, he does touch upon some import-ant reasons why the causality characteristic of mechanical systems is not applicable. First, causality is a relation between events or facts. But the identity of events or facts is always multiple, depending on con-text and interest. Thus no causal connection can take shape without conceptual choices discriminating relevant factors from background conditions. Consequently the causal linkage between representations and the world is not autonomous. Rather, that causal linkage is itself dependent on the intentional. Second, in arguing against Jerry Fodor's causal theory of reference, Putnam rightly points out that no reference can be fixed (thus no reality can be reliably reached) without reflective and normative judgements, which are context- and interest-dependent, and social in nature.

But it is just here, on the issue of unity and diversity of nature, that the ontological approach to explanation has the advantage. In terms of entity, the ontological approach agrees with reductionism that there is not an unbridgeable gap between entities at different levels that cannot be closed by causal connections. Entities at a higher level are always causally explainable by the behaviour of entities at an underlying level. Thus the unity of nature can be argued on the basis of causal connectedness of entities at adjacent levels and thus at all levels.

In terms of laws and principles, however, the ontological approach agrees with anti-reductionism that laws and principles have their roots in the ontology of their own domain (with novel features that are absent in the ontology of other domains at a lower level). Thus they are specific to their own domain and cannot be explained by entities, laws, and principles of other domains at a lower level. Since laws and principles play a direct role in explanation, their autonomous status gives the ultimate reason why explanation is non-transitive across levels. Thus the diversity of nature can be argued on the basis of the emergence of new entities, whose behaviour is governed by laws and principles that are independent of whatever occurs at lower levels.

To illustrate this point, consider once again the emergence of the intentional mind within the physical world.

Putnam criticizes reductionism for having missed the most important features of the intentional mind, namely: (i) being active and constructive; (ii) being potential, as open possibility toward the future; and (iii) being semantic, intentional, and social, searching for meaning, truths, and reassurance from other minds. Yet as Walter Freeman (1995) has shown, it is by no means necessary to retreat into mind–matter dualism in order to account for these features.

Taking humans who have a dynamic relationship with the environ-ment as the primary entities for the study of intentionality, Freeman developed a dynamic model for the objective side of the intentionality structure, which explains why the above features obtain. With this ontological commitment, Freeman's starting point is humans' social and practical action in the environment. First, the metabolic needs of brains for building materials and fuel, which must be searched and found in the world, dictate that the basic behaviour of humans is movement directed into the world, and this requires an internal struc-ture of mind adequate for orientation in space and time. This require-ment is realized in the natural history of humanity by the interactions of neurones with each other to form assemblies, of assemblies to form brains, and of brains to form societies – with the aim of muscling in on a shared environment. Note that here the requirement and its real-ization can only be explained in functional and evolutionary, rather than causal, terms.

Second, the transformation of activity of microscopic neurones to populations is described, tentatively, by non-linear dynamics, as the formation of patterns of neurons' chaotic activity. Here, the micro-scopic neurons contribute to the macroscopic order, and then are enslaved by that order.

Third, Freeman, based on his experimental discoveries – and this seems to be the most important part of his story – tells us that firings go from the patterns of neuronal activity into the motor systems that induce search movements. But they go also as corollary discharges to all of the sensory cortices to prepare them for the consequences of the intended actions, and to specify the classes of stimuli that are sought. Thus intentional perception appears with the emergent patterns of activities in the brain.

Thereby, neurologically, intentionality can be described as the ongoing dynamic construction of motor patterns with concomitant transmission of discharges to all sensory systems in order to update them with respect to the expected consequences. From this notion of pattern construction, most essential features of the objective side of intentionality can be derived.

Whether or not Freeman is right, his account of mind and intentionality demonstrates the entire feasibility of naturalistic theories of those phenomena satisfying the unity of nature, and, more particularly, satisfying the following two general desiderata. First, although the mind emerged from the cosmic evolution of matter, mind cannot be reduced to matter, and the subjective aspect of the brain's neuroactivity cannot be reduced to its objective aspect; thus, (i) the objective aspect of mental activity is scientifically accessible, and can be naturalistically investigated and explained; (ii) the content of the mental is conditioned by the physical and social environment, and derivable by the practical actions of humans as a physical existence within the physical world; and (iii) there are effective interactions between the mental and the physical. Second, the emergence of new, non-primary, yet irreducible ontologies – most important among them the mind – reveals the openness of nature. That is, nature gives more and more surprises and becomes richer and richer in its content. This openness gives ground for diversity, but this should not be interpreted as an indication that nature has become fragmented.

Can we explain the genesis of the universe?

Scientifically, the question concerning the origin of the universe (cosmogony) makes sense only within the framework of classical big bang cosmology (BBC). Supported by empirical evidence, BBC postulates that the universe as a whole has only a finite age and thus an origin. If BBC is a complete account of cosmogony, however, no naturalistic explanation for the genesis of the universe – explanation in terms of what exists in the world – could escape being self-defeating. Since whatever is physically imaginable is physically connected with our observed universe (namely all possible events that are compatible with the observational evidences that we have obtained hitherto), then nothing which is physically imaginable could be responsible for the genesis of the universe.

But BBC is by no means a complete theory, although it has provided a reasonably good account of the evolution of the observed universe. It is incomplete because, first, the evidential support of BBC does not foreclose (although BBC itself does) physical scenarios in which events prior to the big bang, itself the beginning of the observed expansion of the universe, existed and led to the big bang.

Second, it is not even a self-consistent theory. BBC is derived from general relativity, yet the very existence of the big bang in BBC, which is incompatible with any temporal predecessor, breaks general covariance. The inconsistency may be avoided by excluding the mathematical singularity at the very moment of the big bang as not being a physical event or an actual moment in time. Yet the theory thus obtained would be big bang cosmology without a big bang, having accepted self-contradiction in exchange for not being in conflict with general relativity. Lacking a physical big bang, this relativistically bona fide theory has lost one of its merits, namely the cause for the subsequent expansion and evolution.

Third, and most important, when the time following the instant of the big bang is less than the Planck time, the quantum effects are dominant. Thus a proper framework for describing the early universe would be the still-desired quantum theory of gravity rather than the classical theory of general relativity from which BBC is derived.

Once BBC's status as a complete theory of cosmogony is rejected, a vast space is created for naturalistic explanations of the origin of the observed universe. In fact, with the collapse of BBC, the legitimacy of any discourse about the origin of the universe in the sense of BBC is definitely lost. Reference to the big bang makes sense only as a shorthand for a set of events that are the cause of the subsequent evolution of our physical universe as we have observed it – a cause previously and inappropriately designated as the big bang. If we understand the origin this way, then it can be explained by currently accepted physical theories without appealing to any supernatural cause. One of the leading candidates in this endeavour is quantum cosmogony.

The upshot of quantum cosmogony is that our universe originated from quantum vacuum fluctuations of a background space. The physically interesting part of the inquiry is, within the general framework of quantum theory, to find a physical mechanism for quantum vacuum fluctuations to occur – it is suggested to be quantum tunnelling – and to supply a scenario – provided by so-called inflationary theory – that describes how the supposed quantum vacuum fluctuations led to the observed (or, more properly, the inferred) expansion. In the sense of a legitimate cosmogonic discourse, this is the kind of scientific explanation we can reasonably hope for.

But there is a difficulty. Quantum vacuum fluctuations require a background space whose specific properties make fluctuations possible. Where, then, did the background space come from? One way to bypass this difficulty is to explain the origin in terms of quantum emergence from nothing rather than from a background space. Since nothing has the least possible connection with the observed universe, that is, it has nothing to do with any physical entities and properties, such as particles, fields, energy, momentum, or even any classical space-time substratum, except for being susceptible to quantum fluctuations that are justified by the uncertainty principle of quantum mechanics, we seem to have an ontological basis for explaining the origin of the observed universe in terms of a natural cause, that is, the quantum nothing.

Note that the quantum nothing is ontologically quite different from the quantum vacuum. The quantum vacuum fluctuates in the sense that it is the seat of violation of the conservation of energy, and energy is regarded as the defining property of physical substance. Thus the very existence of quantum vacuum fluctuations indicates that the quantum vacuum is a new form of physical substance revealed by quantum physics. But whereas the quantum vacuum resides in a previously existing substratum of space-time, the quantum nothing presupposes no such substratum, and thus appears to be an even newer form of physical substance, revealed by quantum cosmogony. The notion of the quantum nothing is of great importance. First, it provides a genuine case of Spinoza's notion of substance in the sense that it is its own cause of its very existence and of its change, without requiring any further external cause. Since in the cosmogonic discourse any appeal to external cause is either self-defeating or a step to supernatural explanation, the quantum nothing as a self-caused substance has provided a new possibility.

Second, together with the most general principles and laws of quantum theory, the quantum nothing gives an ontological basis for a picture whose coherence and articulateness yield a clear understanding concerning the origin and evolution of the universe. Thus, the quantum nothing is not just another vague and putative something, but has quite specific ontological features and explanatory power, and so deserves a distinct place in our physical and philosophical discourses of cosmogony.

It should be stressed that the naturalistic explanation of the genesis of the universe, grounded on the notion of the quantum nothing, offers a new type of causal explanation. Traditionally, a causal explanation in physical sciences means providing a causal mechanism for the evolution of a system, which is governed by laws under given initial conditions. Initial conditions are thus in some sense antecedent causes. While it is quite legitimate in other cases to start out scientific explanation by specifying initial conditions, in cosmogonic discourse it would be question-begging. The notion of the self-caused quantum nothing escapes this difficulty. Thus, as the only theoretical resource for a naturalistic explanation of the origin, it would also be sufficient for the task. A specific physical model that embodies this understanding of explanation is Stephen Hawking's 'No boundary proposal' (Hawking 1988).

To sum up, the quantum fluctuations of nothing are a legitimate possibility for the genesis of the observed universe; the tunnelling proposal and the inflation scenario have provided more specific mechanisms for the emergence of space-time and matter; together, they have provided an explanation for the genesis of the observed universe. Thus, the move from BBC to quantum cosmogony is a progressive step in our understanding of the universe. Of course, there remain many questions in quantum cosmogony. Yet we have to remember that scientific explanation is a stepwise enterprise. The fact that an explanation of a puzzle always creates another puzzle to be explained does not disqualify it as an authentic explanation. Only with such an attitude can our inquiry about an explanation of the origin of the universe have genuine scientific interest.

A concluding remark

The most important implication of the ontological approach to explanation is this: whenever we have something important but difficult to understand, we should focus our attention on finding what the primary entities are in the domain under investigation. Discovering these entities and their intrinsic and structural properties, rather than manipulating uninterpreted or ill-interpreted mathematical symbols, or speculating on free-floating universal laws and principles, is the real work of science. Mathematical formalisms and universal laws and principles are relevant and important only when they have a firm ontological basis.

Acknowledgements

This work is financially supported by a grant from the Mrs L. D. Rope Third Charitable Settlement, which is greatly appreciated. I am also grateful to Paul Forman and Robert S. Cohen for stylistic help.

References

Arbib, M. A. and Hesse M. B. (1986). *The construction of reality.* Cambridge University Press.

Bell, J. S. (1987). *Speakable and unspeakable in quantum mechanics,* pp. 1–21. Cambridge University Press.

Cao, T. Y. (1998). 'Monism, but not through reductionism'. In *Philosophy of nature: the human dimension* (ed. R. S. Cohen and A. I. Tauber), pp. 39–51. Kluwer, Dordrecht/Boston/London.

Cao, T. Y. (2001). 'Prerequisites for a consistent framework of quantum gravity'. *Studies in History and Philosophy of Modern Physics,* **32** (2), 181–204.

Freeman W. (1995). *Societies of brains.* Lawrence Erlbaum, Mahwah, NJ.

Hawking, S. W. (1988). *A brief history of time.* Bantam, New York.

Hempel, C. G. (1965). *Aspects of scientific explanation.* Free Press, New York.

Hesse, M. B. (1963). *Models and analogies in science.* Sheed and Ward, London.

Hesse, M. B. (1965). 'The explanatory function of metaphor'. In *Logic, methodology and philosophy of science* (ed. Y. Bar-Hillel), pp. 249–59. North-Holland, Amsterdam.

Mellor, D. H. (1995). *The facts of causation.* Routledge, London.

Putnam, H. (1992). Renewing philosophy, Chapters 1–3. Harvard University Press, Cambridge, MA.

Putnam, H. (1994). *Words and Life,* Chapters 1–3 and 20–24. Harvard University Press, Cambridge, MA.

Radicati, L. A. (1984). 'Chaos and cosmos'. In *Particle physics* (ed. I. Bars, A. Chodos, and C.-H. Tze), pp. 33–45. Plenum, New York.

Salmon, W. C. (1984). *Scientific explanation and the causal structure of the world.* Princeton University Press.

Watkins, J. (1984). *Science and scepticism.* Princeton University Press.

Weinberg, S. (1995). 'Reductionism redux'. *New York Review of Books,* 5 October, 39–42.

Jack Goody

From explanation to interpretation in social anthropology

LET me begin by posing the question of the difference between explanation and interpretation. Explanation attempts to answer the 'why' question. That seems to me a general question posed in all human societies as a result of the interaction between language-using animals and their environment; answers differ, some we call myth, others history, others explanation, but the questioning remains. Interpretation involves finding an alternative way of putting things, often attempting to interpret events from the standpoint of the actors. A whole school of sociology, that associated with the name of the great German sociologist, Max Weber, and called by him *verstehen*, claims to be involved in 'interpretative sociology', in understanding what the actor meant, in intentionality, rather than in seeking explanations (even as a preliminary). There was, if you like, a shift from science to ethnoscience, validated in its own terms.

In the social sciences, the opposition between these approaches has turned on the appropriateness or not of established procedures of scientific explanation. Were these possible in the social sciences, where one was dealing with culture rather than nature, with human beings whose possession of consciousness, of free will, reflexivity, it was argued, made deterministic predictions of behaviour impossible (for example, in elections and stock markets). That was an argument elaborated from Weber by philosophers like Peter Winch and adopted by many social scientists themselves, who opted for the emic rather than etic, for the actor's rather than the observer's view, for the subjective rather than the objective.

I am speaking here of social or cultural anthropology rather than these fields known earlier by the same name, archaeology and human biology, which in general utilize accepted forms of explanation. Sociocultural anthropology has long had a problem of changing not only its explanation but its very mode of understanding, resulting in a lack of continuity, a rejection of earlier results, not so much in a Kuhnian paradigm shift as of total change of mode. There has been a constant query, not so much of the question as of the nature of the question.

Let me turn to the story of those changes. Before positivism the use of scientific models became (for many) a word of abuse, of rejection. Anthropological explanations are basically not really different from sociological ones. But they have a somewhat different history. In the nineteenth century, anthropology was very much concerned with the long view of human history, going back to the emergence of humankind, linked with prehistory and human evolution. Its practitioners were interested in human behaviour, especially in standardized actions, in origins. The difficulty was that, while they had archaeological evidence of a material kind, stones, bones, pots, from earliest times, they had no other residue of how humans behaved, of how they married, for example. Consequently the search for such origins was not truly historical but rather pseudo-historical, speculative, and subjected to highly imaginary 'evolutionary' sequencing.

Take for example the practice, a custom, that obtained in Europe until recently, and obtains even today in some circles, of carrying the bride over the threshold of the house into which she has moved. Anthropologists in the nineteenth century, who worked from books rather than from nature, from observation, interpreted this custom as a *survival* of yet earlier forms of marriage, as a kind of cultural 'memory trace'. They posited an earlier stage of society in which people practised 'marriage by capture'. That is, in which groups of men, who were prohibited from marrying their own sisters by the incest taboo, set out to capture women from other groups as brides, a kind of generalized Rape of the Sabines.

That phase of human life was supposed to be patrilineal and patriarchal, with the dominance of men, and to have followed first one of 'primitive promiscuity' when there was no incest taboo, then one of matrilyny or matriarchy, when only the mother's line was certain. The story was elaborated by an American lawyer–anthropologist, Lewis H. Morgan, whose book *Ancient society* had a great influence on Marx and Engels (especially in 'The origin of the family' – once again origins) and through them the whole of Soviet anthropology and long-term cultural history.

Ideologically we can see this being set up as the polar opposite of the monogamous marriage of nineteenth century Europe, to which human society 'evolved' by a process of social evolution, from promiscuity to monogamy.

The problem about these schemes was that there was not the slightest evidence for them, and later on, when different forms of explanation come to the fore, they gave historical and evolutionary explanations a bad name, especially when they consisted of this unilineal evolutionary kind, assuming one single way in which humans developed in society.

The change came at the beginning of the twentieth century, and it came from two sources, from sociological theory in the works of the French sociologist, Emile Durkheim, who was very interested in other cultures, and from the shift from armchair anthropology to observation, to fieldwork.

Fieldwork in this country could be said to have begun with the famous Cambridge Torres Straits expedition of 1898, which took place largely under psychological, medical, and biological auspices. In the main they were interested not in history but in experimental observations and the correlations they might throw up – the relationship between two sets of data, a comparison of social facts as Durkheim called them. The 'subject', too, changed; the previous focus was obviously on earlier and primitive societies. However, with observations in the contemporary world, all societies were available; even earlier ones could be reconstructed by comparison with contemporary hunters or gatherers. Clearly that possibility, too, changed with changing times; so did the subject leading to some present preconceptions which tend to reject all attempts at explanation.

But the major shift came from the work of two anthropologists who went off to the field by themselves, not in a team, to try to understand the life and behaviour of the simpler human societies. One of these was Bronisław Malinowski, a Polish anthropologist working from the LSE, who went to the Trobriand Islands in Papua New Guinea; the other was A. R. Radcliffe-Brown, who worked from Trinity College, Cambridge and went to the Andaman Islands in the Bay of Bengal to study a community of hunters and gatherers who were pygmoid in physical type.

The most innovative as far as fieldwork was concerned was undoubtedly Malinowski. He spent some four years in the Trobriands during the First World War, partly because of his enemy status; he learned the language and produced the classical ethnographic account of a matrilineal people, which he used to try and modify Freudian theories of the universality of the Oedipus complex and the pseudo-historical *Totem and Taboo* (1912), just as he used data on the exchange system of the Kula ring, a ring in which necklaces went in one direction and armbands in another, to challenge certain theories of classical economics.

Malinowski elaborated the technique of intensive fieldwork. His explanations were therefore not historical in terms of origins, but used either limited comparison, as with the Oedipus complex, or, more significantly, were *'functional'* in terms of what a custom contributed to the total running of society. The model was in a sense biological. The role of the heart, the function of the heart, the explanation of the heart, is what it contributes to the running of the body, a question of the part and the whole. One asked contemporary questions about what customs, institutions, actions, *did* in the context in which one found them. One asked questions about the *structure* of the situation, the nature of social relationships. It was a socio-functional explanation which achieved its sociological expression in the writing of Robert Merton and its philosophical justification at the hands of Ernest Nagel. Since time had been mishandled, it was now not handled at all, excluded, but functional anthropologists spent some effort in trying to bring it back, at least in the form of cycles (for example, of domestic groups) or of observations of the same community at different periods.

Let me return to the carrying of the bride over the threshold. One in fact sees somewhat similar customs in many societies in Africa today. Among the LoDagaa of Northern Ghana, where I worked for a number of years, a new bride was accompanied to her husband's house by her own kin and by her husband's. As she went, she cried like a child and some outsiders thought this represented a forced marriage of the utmost cruelty. But if you looked at the actor's point of view, which is what social anthropologists are supposedly trained to do, things appeared very different. A girl has to cry when she leaves her natal home, people explained, if not it would be like a bitch going to a dog; she would show no sign of regret at leaving her parents' home. So this custom represented the cultural ambivalence, the tensions that existed when a woman transferred her main attachment from the house of her parents to that of her husband. It had nothing to do with origins in some earlier system of marriage but represented the situation as it existed in the present. The social function of the custom was to express the tension that existed in the physical transfer of the bride. If it was in any sense a *survival*, it had to be understood in terms of its current role.

This shift in perspective, largely due to the immersion of the investigator in a particular society rather than in other people's books, the learning of the language, the ability to get close to the actor's point of view, led to a concentration on socio-functional and structural explanations. History was out, not necessarily bunk as Henry Ford is reputed to have said, but out. Even more so evolutionary explanations that had proved so powerful in the hands of Charles Darwin and his followers, as in archaeology and human biology. Instead we arrived at functional explanation. That led to a general shift, though not really a paradigm shift in a Kuhnian sense.

As I have indicated, functional explanations were made partly on a biological analogy; the contribution of the part to the whole living organism. At times Malinowski took this further and tried to relate the human institutions surrounding food to the biological drive of hunger. That type of bio-functional explanation did not prove very profitable because of the great gap between 'cause' and 'effect', and the considerable variety of those effects in practice. The explanation was over-generalized (under determined), covering too many possibilities.

Yet that type of vague biological explanation was also adopted by many others, and continues to be endorsed in some branches of anthropology influenced by biology or sociobiology, for example even by Emile Durkheim when on one occasion he attributed incest to a *drive* for incest, a singularly circular form of non-explanation. We cannot disregard the biological but reductionism even in the form of socio-biology has not proved to be a very powerful tool at all.

In the main Durkheim was the theoretical sponsor of socio-functionalist/structuralist explanations. He was concerned with the analysis of what he called social facts, and although he occasionally resorted to biological explanations of pseudo-historical ones (though reasonably well founded), his main mode was functional. An example of this pseudo-historical mode was his proposed progression from societies based on what he called mechanical solidarity to those on organic solidarity, that is from societies in which all the units (for example, clans) were similar, and who were bound together by identity, and those more complex societies with an advanced division of labour (see Adam Smith) where the activity of each unit depended upon the complementary activity of the other, for example, car-makers or road-builders, and who were therefore related *organically* like the parts of a body.

But his more usual mode was functional, for example, in his account of the corroboree ceremonies of the Arunta of Australia (using the work of other observers, such as Spencer and Gillen), which are held when the tribe gets together, dances, sings, renews social ties between the members. Durkheim attributes the efficiency of the rite to the 'effervescence' aroused in the participants, which marks their collective relationships and which helps to hold the group together. This is the function of the rites and in this sense the explanation for them. He is not concerned with historical origins, with diffusion, or borrowing from neighbouring peoples (questions which he might have seen as rather mechanical), but with the part they played in the social system as a whole. The present not the past, which is how he would explain so-called survivals.

That was not the only kind of functional explanation provided by Durkheim. In a well-known study entitled *Suicide*, he compares rates in a variety of Western European countries. He worked specifically with those from different Swiss cantons, which enabled him, for example, to compare rates in ones that were dominantly Catholic with those that were dominantly Protestant. He noted that rates were higher in the latter and concluded that it was because they were 'forced to be free', had to work things out more for themselves, and were hence less certain of life.

Suicide was therefore a function of solidarity in a mathematical sense, where X is a function of Y. He comes to his conclusion by means of a critical experiment comparing two groups with regard to the same variable. He does the same for married versus single, the latter having higher rates. A similar procedure had been recommended by Francis Bacon in the sixteenth century for the then new (natural) sciences. Here it is not so much the crucial experiment (which one can rarely use in the social sciences where one is dealing with human actions in concrete situations) as the crucial comparison; the comparative method substitutes for the experimental.

Durkheim's study of suicide is an impressive piece of work that has always for me been an important model of explanation, especially in comparative work. There are two problems, however, that in my view raise general points on the question of explanation. The first is that the independent variable, solidarity, is somewhat too vague and all-embracing; it can include so much and therefore resembles some explanatory variables used by historians, especially historians of the *mentalités* school, for example when they talk, as do Ariès and others, about the increase in affection (love) for children in the sixteenth century. Secondly, these particular explanations are psychological in character. Some people, including Durkheim, object to shifting levels of explanation, for example moving from the sociological to the psychological, or the social to the biological, but that is not my objection and I do not think we should altogether exclude such shifts of level. But they can easily be made prematurely and there is a problem in interiorizing explanation in this manner because, desirable as it may be to probe the human psyche, it is unclear what one is doing when dealing with solidarity or affection. Undoubtedly they are to us recognizable aspects of social facts, but they are difficult to assess and represent in the same way as suicide rates. So recourse to such internal features, perhaps hypothetical, may obscure rather than reveal.

The third point is this. Types of explanation have to be linked to the nature of the enquiry. The analysis of the data gathered from intensive fieldwork is quite a different project than that involved in a comparative analysis of data from several groups, or indeed from a sample of worldwide societies such as exists in the *World Ethnographic Atlas*, begun by the American anthropologist, G. P. Murdock, to enable people to undertake cross-cultural comparisons of a global kind. Or indeed a project in comparative history over the long term.

The analysis of field material that you or others have gathered from a particular society promotes functional or structural explanations because one is concerned with how that society works. It is of course an advantage to have comparative and historical perspectives in mind. I would suggest that this is true of all the social sciences. Much European history is diminished by the assumption that only Europe has taken a course that contributed towards modernization, for example, when a look around the globe would have shown that the assumption was far from the case. This is true of the assumption made by historians such as Stone, or sociologists such as Giddens, and indeed many others, that love, at least in the form of romantic love, was confined to Western Europe from the twelfth to eighteenth centuries.

Let me return specifically to the analysis of data gathered intensively from one society. At one level, as I have suggested, explanations are almost bound to be phrased in functional terms, what a particular custom or institution contributes (or fails to contribute) to the working of society, to the social system. And to establish these, Malinowski and Radcliffe-Brown, like Durkheim, set aside and dismissed the pseudo-historical explanations – of marriage customs, for example.

But as I have indicated, they declared themselves uninterested in historical explanation altogether. In doing so they seem to have unnecessarily reduced the range of problem they could deal with. For instance, the LoDagaa were in effect a completely oral society before the advent of the Europeans at the beginning of the twentieth century. They knew in a very general way about Arabic writing and even made use of some written Arabic 'charms' (*safi*) for therapeutic and similar purpose. Some time after the Europeans came, at the end of the 1940s, a government primary school was established in the area where I worked and the pupils learnt to read and write. The effects of this (historical) change were widely felt.

1. Minutes of meetings were taken, leading to the formality of following-up decisions on the next occasion.

2. Funeral gifts were recorded, like Christmas cards, leading to a much more precise concept of reciprocity, or morality.

3. Those who went to school became dispersed spatially and in the jobs they did, part of a national, indeed worldwide economy.

4. Agriculture, at least with the hoe, became a devalued occupation.

These diachronic, developmental changes were not founded on unprovable hypotheses of long-term evolutionary processes, on pseudo-history, but had to be explained by examining concrete data concerning the recent past. Types of developmental (historical) explanation became more common in social anthropology as its practitioners

(1) became increasingly interested in comparing past and present states of one society

    (a) either by returning to do further fieldwork after ten or twenty years (repeated observations by the same individual);

    (b) by examining earlier documents which for Africa were now becoming more available in the archives;

(2) returned to look at earlier long-term hypotheses in the light of new field data (as for example in the work influence by Marx);

(3) collaborated with historians or with history, as in the case of Macfarlane and others;

(4) focussed on contemporary problems of social change which all simple societies were undergoing as the result of colonization (Africa), European emigration (USA), and specific development projects which now attracted funds for research.

Of course diachronic explanations had never entirely disappeared but now they were given renewed legitimacy.

There was another kind of explanation that became common in social anthropology and had a considerable influence on the rest of the social sciences, especially those that shied away from historical explanations (perhaps as being insufficiently 'theoretical', too *événementielle*), and who saw structural functionalism as being too vague and indeterminate and the alternative form of 'mathematical' functionalism, the analysis of variables, as being too positivistic, too scientific, adapted to the natural sciences but not to the social ones. Humans were different; they have consciousness, variability, and cannot, they suggested, be analysed in determinate ways. So that notions of cause and effect are inappropriate; humans obey no laws. That was the objection of some fieldworkers to the claims of Radcliffe-Brown and others. Instead they promote notions such as 'elective affinity' (Max Weber – note the stress on choice) or on structural compatibility.

Some of these objections are based on the mistaken idea that cause and effect means that one assigns one cause (X) to a certain effect (Y); that is, that it involves a notion of single-factor determination. My work on the effects of literacy on social actions has been criticized in this way. This is a crude comment. It is true that not infrequently in the social sciences one is putting forward an explanation of a causal type in a single-stranded way because of the nature of (written) discourse, which takes one factor lineally at a time (for example, the rise of capitalism is attributed to the Protestant ethic). But any adequate causal sequence must allow for a plurality of factors, of independent variables, and it is even more difficult to make an adequate assessment of the role for each one. In effect that can only be done by giving a numerical value to each variable, that is, 60% to the Protestant Ethic, 20% to scientific/technological developments. Following such an allocation (carried out by some acceptable measuring procedures), there are statistical techniques, known as path analysis, which enable one to calculate Beta-weights and to work out a multifactorial causal diagram. It is only rarely, for example through the *World Ethnographic Atlas*, that we can engage in such procedures.

An alternative method adopted by a group of anthropologists ('structuralists') was associated with the name of Lévi-Strauss, who was much influenced by Durkheim (and more specifically by his sister's son, Marcel Mauss) but whose specific inspiration came from structural linguistics. Explanation consisted in a search for 'structure', which referred to some underlying (unconscious) pattern, unlike the more surface notion of structure of the structural-functionalists.

Basic to their notion were linguistic patterns such as the consonantal pattern

d	p	f	voiced
t	b	v	unvoiced

and the vocal triangle

$$i$$
$$e \qquad o$$
$$a \qquad\qquad u$$

Following the lead of the Prague school of linguists, Lévi-Strauss looked for homologies of these, that is, repeated patternings in difference social spheres, for example, the Cooking Triangle. Or more pervasively applicable, the table of opposites as in the consonantal table. For instance, anthropologists engaged in the search for 'symbolic' opposites and equivalences (polarities and analogies) claimed to discover lists like the following:

White	Black
Day	Night
Right	Left

Male	Female
Sun	Moon
Good	Evil

The notion was that one should be able to formulate a set of symbolic oppositions and equivalences for each society, which would correspond, *by homology*, with similar tables for other domains and this totality would represent the long-term structure of society, a structure that was specific to that society.

One major problem with these tables was that they differed little from society to society and therefore seemed to be saying something not about the structure of specific societies but rather of human thought itself. That commonality has led to an interest by cognitive anthropologists and, following a Chomskian model whereby language capacity was built in, they have enquired what else – for example, category systems – are built in to the human psyche, to *'l'esprit humain'*, and so are not in this sense cultural (learnt) but inherited. A lead in this enterprise has been given by Sperber and his collaborators in France.

However, there is another problem with these tables that is of a cultural kind and that relates to their rigidity. They do not allow for alternative usages and therefore raise what I call, 'The Black is Beautiful' problem. It is inconceivable that in Africa black at times is not also positive as well as negative, that it represents rain, fertility as well as night, witchcraft.

In my view such constructs (which I see as essentially literate) cannot be held to explain very much, although Lévi-Strauss has made imaginative use of them in his analysis of myths, *Mythologiques*. All is reduced to simplification, following the lines of Russian analyses of folk tales where the stories are reduced to boy wants girl, etc. Earlier, a similar kind of approach was applied to the structural analysis of Shakespearean texts, by Wilson Knight and Caroline Spurgeon. Once again, all was reduced to an opposition, for example, between Order and Disorder. That binarism reduced to nothing all the staggering complexity of Shakespeare's verse and imagery.

Admittedly reductions of this kind are a kind of explanation. They give comfort to many who then think, when it is reduced to a simple formula, that they have 'understood' the play or the culture. Many scientific understandings are of this kind and can be very generative – take the periodic table. But there must be criteria of relevance, of applicability, of evidence and proof: any table will not do. Or take Talcott Parsons, and the four square table to which he tried to reduce complex social phenomena. This approach looked like analysis and provided sociologists with a measure of control; it became very popular. But subsequent experience has shown that control to be illusory and the procedure has been abandoned without, I think, bearing any fruit.

Finally there has been a further shift, moving from explanation to interpretation, from the etic to the emic, in the form of post-modernism. This movement has rejected the binarism of the structuralist approach, rejected anything that might smack of the scientific as 'positivistic', and indeed rejected much of the explanatory thrust provided by the Enlightenment. It has insisted on the value of the emic, of understanding the actor's viewpoint. As I have noted, all social anthropological enquiry does that, as does Weberian sociology. But whereas the latter has seen such comprehension as a phase in the explanatory process, post-modernists have wished to stop there, insisting that that is all one can know. The limitations of explanation have been linked to the stress they have placed on distortions in knowledge and on trying to understand the interaction between actor and observer, in particular understanding the personal standpoint ('prejudices') of the observing individual, and especially of themselves.

Such enquiry becomes self-reflective to a high degree. The story is told of the anthropologist interviewing a Navaho Indian. After an hour's work the Indian turns round to the anthropologist and asks,

'And when are we going to talk about me?' Such reflexivity has some positive sides. It leads to questioning how an observer's experience may have qualified his observations, or what influence the fact that Malinowski saw himself as a Polish nobleman and Radcliffe-Brown as of English yeoman stock had on their work. Such questions are useful, but they affect all scholarly work, not simply anthropology (though that is perhaps more vulnerable simply because it attempts to interpret the actor's point of view). We need to find ways of coping with any such bias. But post-modernism has tended to see this potentiality as devaluing anthropological observation in itself, leading in the end to a rejection of explanation in the social sciences, which are diminished by being 'social constructs'. That is an argument that has been comprehensively refuted by Randall Collins in a recent book on the sociological analysis of philosophy.

Looking at the practice of social anthropologists, and of some other social sciences over the last century, we see a general tendency to move from explanation to interpretation, partly on the grounds that the human sciences are different. Certainly we need to try to understand the actor's point of view, the meaning of social action to him or herself. That is the essence of modern social anthropology, arrived at through language learning and participant observation. But for me that remains a preliminary level and we should continue to try to look for forms of explanation that are relevant in other spheres, functional, structural, causal if possible.

So I want to return to my title, 'From explanation to interpretation in social anthropology', and add the phrase 'And back again?', partly as a reflection of what I myself try to do, partly as a prayer for the future of the social sciences. We can advance only if we consider problems in which the rest of the scholarly world is interested and if we go for a plurality of explanatory forms.

Jon Turney

12

Passing it on:
redescribing scientific explanation

> Common metaphors are the best because they
> are the only true ones.
>
> JORGE LUIS BORGES

> A new kind of physical entity, a relativistic,
> quantum-mechanical kind of string.
>
> STEVEN WEINBERG

SCIENTISTS often describe that eureka moment. The struggle to
fathom some phenomenon, or to extract meaning from some experimental results, is suddenly rewarded with a flash of understanding.
Common to many such accounts is the idea that part of the pleasure
lies in seeing something no one else has seen before. But that benefit
carries a price: then the newly enlightened one has to explain it all to
everyone else.

That explanation is hard enough when it is directed at fellow pro-
fessionals. If there is a radical reconceptualization involved there will,
of course, be scope for numerous misunderstandings, for uncompre-
hending rejection as well as glad acknowledgement. But let that pass,
even though that means sidestepping a great deal of the history and

philosophy of science. A further set of obstacles appears when new scientific notions come to be explained to non-scientists, when they shift from what Ludwig Fleck called the esoteric to the exoteric realm (Fleck 1935).

It is pretty obvious what some of them are. You never know what non-experts may know, but it is best to assume that they have little formal knowledge of the problems under discussion, little taste for the jargon of the field. They will almost certainly not have the training or the inclination to follow any advanced mathematics. And their stock of examples of how things in the world work, of mental models – items you can make use of to explain what something new is *like* – will probably overlap only partially with that of a trained scientist.

At the same time, non-specialists' appetite for new scientific explanations is impressive. And there are plenty of people trying to satisfy it, as the often remarked boom in popular science publishing of the last 15 years or so testifies. Just look in any large bookstore and you will probably be convinced that there has never been so much science explained to so many people by so many clever authors. Yet when you examine such books, it can be hard to find a satisfying way to describe what is going on. In particular, it is not obvious what counts as a good explanation of some scientific subject. If anything is distinctive about popular science books as a type, it must surely be that their authors have to explain unfamiliar ideas intelligibly. And they draw our attention to an important feature of explanation. While it is fascinating to discuss good explanations in the abstract, in practice explanation most often involves somebody explaining something to somebody else. A fair test of whether the recipient has understood is then whether they can explain it to somebody else in turn.

So how should we go about this cultural work? Unfortunately, just as there is no full account of explanation, philosophically speaking, so there is no tried and tested protocol for producing explanatory text. There is often critical agreement that some writers (Richard Dawkins, say) are spectacularly good at it. There are appealing slogans like Jerome Bruner's assurance that anything can be explained to anybody with a 'courteous translation'. There are examples of how to develop and test such a translation, like Russell Stannard's account of how he wrote his splendid *Uncle Albert* books to try and explain twentieth-century physics to young readers (Stannard 1999). But beyond that, it is not obvious what vocabulary we might draw on to describe how this kind of explanation is done, or to begin to evaluate when it is well done.

Stannard's beguiling fictions immediately suggest a place to look, though. First and foremost, they are stories. And it is certainly plausible to claim that, whatever the prevalent style of explanation in a particular academic discipline, when an explanation is conveyed to others, outside the discipline, it will be cast in the form of a story. Putting it that generally immediately demands clarification. That is partly because popular science can involve so many different kinds of story-telling. I do not mean here, for example, the kinds of stories which publishers look for – stories about, for example, 'a Lone Genius Who solved the Greatest Scientific Problem of His Time'. As the blurb for Dava Sobel's deservedly successful *Longitude* goes on to say, she tells 'the dramatic human story of an epic scientific quest' (Sobel 1996). This kind of narrative is a fine way of engaging the reader (though not a particularly persuasive take on the history of science), but is not the kind of story I have in mind here.

Nor do I mean the kind of narratives which scientists routinely deploy among themselves. The professionally sanctioned rhetoric of science, extensively explored in recent years by literary analysts, includes important aspects of explanation for immediate persuasion of peers. Parts of this are a kind of meta-explanation – accounting for the things left out of the explanation proper. As other chapters (e.g. Chapter 1) of this book point out, there are several kinds of selection which have to be dealt with before a new explanation can carry conviction. They include deciding what features of the world need explaining, which causes matter, and which of several correct explanations are important. If answers to these questions are not contained within the explanation, there must be some larger story which accounts for the selection made under each of these headings.

There may also be a high-level narrative at times of major theory changes. Alisdair MacIntyre has argued that narrative is the key to resolving the problems which the Kuhnian account of science poses for notions of progress. In his view, if a paradigm shift really occurs, it is only accomplished when accompanied by creation of a second-order narrative which explains why the new paradigm is preferable to the old one (MacIntyre 1977). Such narratives are especially likely to deal explicitly with the kinds of selection just referred to. This is something which, if the two are genuinely incommensurable, cannot be done simply from within the new paradigm. Again, this kind of story is often found in popular science books, but this is not the aspect of popular science explanation I want to probe here.

Closer is a narrative which is an inseparable part of some disciplines. The historical sciences – cosmology, geology, evolutionary biology, palaeo-anthropology, and developmental biology – are all committed to narrative in the simplest sense of relating the unfolding of events over time. Since the middle of the twentieth century they have all developed in ways which permit their various timelines to be joined up more or less plausibly to create what the critic Martin Eger has called the new epic of science (Eger 1993). The existence of the new 'grand narrative' is an important part of the appeal of contemporary popular science, to judge from the selection of topics you find in the bookstore.

So all these narratives can be found in works which attempt to explain science to lay readers. Often they are intermingled. *A brief history of time*, for example, combines a narrative of the universe since the big bang with a narrative of the author's own life in science *and* a story of the history of ideas in cosmology which places Hawking in succession to Copernicus, Newton, and Einstein. But still none of these gets to the heart of the explanatory problem.

For that, we need to consider the sense in which any explanation can be figured as a story. The final caveat is that I do not intend here to endorse the post-modern conceit that science is just another form of story-telling, and one story is as good as another. The interest here is precisely in the construction of stories which carry conviction with readers as a fair attempt to describe scientifically based realities – however soundly based they may be – while satisfying scientists that they convey some impression of what (or how) they actually think, even if they adopt quite different language. It is tempting to argue, following the literary critic and cognitive scientist Mark Turner, that they need to be stories because our minds are constructed to understand the world through story (Turner 1996). But it is not necessary for this discussion to go all the way with Turner and those of like mind, merely to accept that stories may be a particularly effective way of making an explanation.

Here, I want briefly to review one general account of explanation as story-telling in a slightly different context, and then see if it can be adapted to help analyse the explanatory work of popular science texts. This scheme for describing explanation was recently developed by John Ogborn and his colleagues at the Institute of Education in London to capture some of the essentials of science teaching (Ogborn *et al.* 1996). Clearly, school-teaching has some crucial features in common with popular science writing, as it involves explaining scientific ideas to non-scientists, or perhaps proto-scientists. Equally, there are important differences, in audience, in motive, and in the resources the explainer can draw on to accomplish their explanatory task. Let me explain in a bit more detail.

In their book, *Explaining science in the classroom*, Ogborn's group tries to combine linguistic and cognitive science approaches to what happens in science classrooms when teachers are struggling to impart understanding of new phenomena. In recounting their analysis of videos and transcripts of a series of teaching sessions in English secondary schools, they see explanation as a special kind of story-telling. As they see it, some vital features of a story are that:

- there is a cast of protagonists, each of which has its own capabilities that are what makes it what it is;

- members of this cast enact one of the many series of events of which they are capable;

- these events have a consequence which follows from the nature of the protagonists and the events they happen to enact.

Perhaps the most important feature of many of the scientific stories which serve as explanation is then the unusual, and hitherto unfamiliar, qualities of the protagonists and their capabilities. Thus, for example,

> An explanation of the mechanism of heredity introduces novel actions of novel entities. A mother and father passing characteristics to their child turns into a story about a molecule, DNA, which can make copies of itself. Possessing blue eyes or brown hair becomes possessing a set of chemically coded sequences in DNA. *The story involves unfamiliar objects which do unfamiliar things in an inaccessible world* (p. 10; emphasis added).

This already sounds promising as a way of thinking about popular science. It highlights one of the most obvious problems for the explanatory writer, the conceptual difficulties to be conveyed. What kinds of things are the scientifically defined entities – atoms or genes – which, normally, no one ever sees? What does it mean to conceive of gravity as a distortion of space-time around a massive object? What are virtual particles, electron tunnelling, evolutionarily stable states of hydrogen bonds? It also emphasizes what you might call the experiential difficulties. Scientific explanations deploy new kinds of entities, which usually exist in realms beyond the normal human senses. Their behaviour typically contradicts intuitions tuned to the middle range of existence. Human beings have direct knowledge only of things of medium size – a few millimetres to a few hundred metres – and which last for a few seconds to a few decades. They can recover information directly (ignoring the mediation of the sensory organs) only through registering light of certain wavelengths, sound of a relatively narrow range of frequencies, and so on. Scientific observation transcends all these limitations at the price of more and more complex mediations between observer and observed.

From here, the book goes on to develop an account of explanation in teaching based on the use of words, gestures, and diverse objects to create stories about the kinds of things there are and the kinds of things they can do. The task of explaining is seen as being one of bringing 'entities of science' into being for students. Their description of how teachers explain has four parts: creating differences, constructing entities, transforming knowledge, and putting meaning into matter.

Creating differences is seen as 'the fundamental motor of communication' between any two participants in a conversation. They may be differences of knowledge, information, interest, or power. For an explanation to be called for, there must be an acknowledged difference in knowledge.

Constructing entities is a matter of populating the explanatory story with things which can behave in a way which accounts for the phenomenon to be explained. As they put it, 'much of the work of explaining ... concerns the resources out of which explanations are later to be constructed. Protagonists have to be described, with what they can do and have done to them, before any story which explains a phenomenon can be told.' We could simply describe this in terms of introducing scientific concepts, but entities is deliberately chosen here as a more general term, covering a very varied set, an 'ontological zoo'. The widely recognized staples of popular science writing, metaphor and analogy, are important here, to help people understand new things in terms of ones which they are already familiar with.

Transforming knowledge means the continuous process of making and remaking ideas, to which the classroom explanation makes one series of contributions. Analogy and metaphor again play key roles here, as does narrative.

Finally, putting meaning into matter is intended to suggest the role of demonstrations, or pupils' own experiments, which are intended to show that the material world really does behave in the ways the ideas and entities which have been brought into play imply.

Ogborn and co suggest that these four sets of ideas offer 'the beginning of a new language for thinking about the act of explaining science in the classroom'. I think they are also useful for analysing the explanatory work being done outside the classroom, in popular science texts. Each of the things which a teacher going through his or her very complex, multi-modal explanatory performance may be doing has its equivalent in the rather more restrictive symbolic world of the science writer.

In text, creating differences will mean trying to establish the differences between what the reader or people in general think is the case, or what is generally taken to be true, and what is the case according to the author. The contrast may be signalled by a range of devices, including imputing views to the reader directly, describing what someone else thinks – where the someone else may be another scientist or a historical actor or someone from another culture – or outlining a 'common sense' view. Sometimes it may involve persuading the reader that what is about to be explained is something that does, in fact, require explaining, as Richard Dawkins strives to do for the existence of design in the living world in the opening pages of *The blind watchmaker*.

Elaborating on constructing entities, Ogborn and his colleagues write: 'A scientific explanation needs to invoke protagonists which are not part of common knowledge. Explaining to someone then requires describing the possible protagonists as well as accounting for what they may have done.' Authors have to find ways of introducing the cast of characters they want to use in their explanation – gene, enzyme, atom, antibody, black hole, neurone, magnetic monopole, and so on – and describing their properties.

Transforming knowledge may be a more subtle matter, and harder to locate precisely in texts because it necessarily requires extended treatments of single topics. Sometimes, one is invited to think of some entity one way, by way of introduction, then later on invited to think of it another way, as a substitute, a supplement or an addition to what was said before. This may often be done through telling a historical story.

Perhaps the most awkward translation is the fourth term. When they discuss putting meaning into matter Ogborn's group are strictly referring to something which happens outside text. They use this phrase to describe the use of demonstrations as a controlled display of a particular kind. The direct textual equivalents of this work in popular science books would be *accounts* of experiments or demonstrations. But we should also examine the use of thought experiments, much more common in popular science than in technical literature, or various devices which try to induce us to see things from the point of view of the entities being described: 'if you were an electron/white blood cell/free radical/cosmic ray ...'.

How does all of this lead to the construction of explanatory stories – which 'tell how something or other comes about'? Elsewhere I have discussed some widely read popular biology texts using this scheme (Turney 2001). Here, I want to sketch a different case study, of a single book in physics, Brian Greene's *The elegant universe*. This ought to be a good test. It is the latest book-length treatment of some of the most difficult, but scientifically most exciting contemporary physics, concerning superstring theories of fundamental forces and particles. The author is himself a leading string theorist. The book has been warmly received by reviewers, including fellow theoretical physicists like George Ellis, who describes it as 'an explanatory tour de force'. It has also been a publishing success, selling hundreds of thousands of copies, according to newspaper reports. And it attracted further plaudits with the award of the UK Aventis Science Book Prize in May 2000.

A further reason for focusing on superstrings is that the ideas are relatively new. Indeed, the theory itself is still not fully formulated. Although there have been popular articles about superstrings since the mid-1980s, there is nothing like the body of writing about them that there is about either relativity theory or quantum mechanics. Earlier accounts of particle theory which feature superstrings, like Steven Weinberg's *Dreams of a final theory*, treated them much more briefly than Greene's. Weinberg's description begins, unpromisingly, with 'these strings can be visualized as tiny one-dimensional rips in the smooth fabric of space' (Weinberg 1993, p. 170). Since one does not have any very clear idea what the fabric of space might look like, this notion of a rip or tear as an aid to visualization seems problematic.

Mixing two metaphors in this fashion (perhaps the strings represent the fabric of space unravelling?) suggests that there is not yet a widely accepted formula for describing these novel mathematically derived entities in plain text. Each of the two earlier revolutions in physics prompted numerous efforts to explain them to non-physicists, efforts which continue today. Although books have individual authors this becomes a collective enterprise, as there often appears to be consensus about the best, or least worst, way to explain certain aspects of a theory. In this way, successful physical theories gradually establish traditions of explanation which new writers can draw on. Greene does just that here in his opening chapters on relativity and quantum mechanics. As usually happens, new scientific explanations are deeply embedded in older ones, and he spends the first hundred or so pages of the book on a largely conventional account of relativity and quantum theory, and of why there is no way of uniting the relativistic framework for understanding gravity with quantum mechanics. Then he unveils superstrings as the solution to the problem.

So let us look more closely at the chapters in the middle of the book, which begin Greene's extended treatment of superstring theory. What follows is not an attempt to explain superstrings, which I am not remotely qualified to do. For that, one should refer to Greene in full. The point is to describe how the explanation is accomplished, with a view to getting some indications of how to do such explanatory work. And I am going to try to describe it using the vocabulary offered by Ogborn and his colleagues.

In their terms, creating differences is fairly straightforward in this case. Greene understands string theory. We do not. He makes us want to understand it, as I have said, by promising that it shows the way to a grand unified theory of physical forces. And he offers the additional promise that it furnishes a far better *explanation* than the so-called Standard Model of forces and particles, in the sense that it should be able to predict all the observed particle masses and force strengths rather than merely record them.

Creating entities is, in one sense, also straightforward. There is only one actual entity involved. The whole point of the theory is that the universe contains just one fundamental kind of thing, which can manifest itself in numerous different ways, as all the particles and forces of conventional theory. But we need to have a non-mathematical sense of what kind of thing it is. How is this done?

The reader already knows, of course, that superstring theory will involve string-like entities. The word appears in the subtitle of the book. The first move to characterize them more fully is to answer the question: What kind of string? A ball of garden twine, perhaps? A puppet-string? A bow-string? No, none of these. The macro-string we are invited to contemplate to help understand the new micro-strings is a musical instrument, with strings vibrating under tension, and generating harmonies. We are, apparently, in a Pythagorean universe.

But no sooner are we dusting off our intuitions about violins and piano strings than the image undergoes some rather violent alterations. For these strings can never be plucked or bowed. They are almost inconceivably small. Strings are (10^{20}) times smaller than an atomic nucleus. Now strings appear as 'tiny, one-dimensional filaments somewhat like infinitely thin rubber bands, vibrating to and fro' (p. 136). Nor are these strings stretched between fixed points like those in a musical instrument because, again like rubber bands, they are looped around. In fact, each of the diverse entities which physicists used to tell us were point particles are made of a single, tiny, oscillating loop of string (p. 140).

Each loop is not made of any ordinary material. In fact, it is not made *of* material at all. It is what material *is*. Just as, once, it made no sense to ask what an elementary particle was made of, there is no real content to the question, What are strings made of? They just are. Rather ask what they can do. What roles can they take in an explanatory story?

The main thing they can do is vibrate, oscillate, or resonate – Greene uses the three terms almost interchangeably. And to explicate them he reverts to his musical image. To understand how string theory has predictive power, he says, think about more familiar strings, such as those on a violin. A musical string generates tones by resonating at frequencies which allow a whole number of waves to occupy the length of the string. We are invited to visualize superstrings in the same way. And, crucially, just as the different vibrational patterns of a violin string give rise to different musical notes, the different vibrational patterns of a fundamental string give rise to different masses and force charges (p. 143). He reiterates this core analogy a few pages later, when he writes that what appear to be different elementary particles are actually different notes on a fundamental string (p. 146).

In this chapter, then, Greene begins to create the string as an entity. So far, it has two essential features. Unlike fundamental particles, as previously conceived, it has extension: it is one-dimensional rather than dimensionless. And it can, in some sense, vibrate. In addition, as explained in detail I will not summarize here, strings are vanishingly small, a consequence of the tension in each string being fantastically high. But at the end of the chapter he introduces a third property, which is radically out of line with our experience and intuition concerning everyday vibrating strings. The theory provides for individual strings to merge or divide. A particle collision is now pictured as the merger of two strings to produce a third. This string travels a bit, and then releases the energy derived from the two initial strings by dissociating into two strings that travel onward (p. 161). But it is not obvious what kind of picture this can be (in spite of the insertion of a diagram at this point). The notion of collision and merger is completely unexpected from the macro-analogy which Greene has placed at the centre of his explanation so far.

So the kind of entity Greene creates to convey 'The essentials of superstring theory', as he titles this chapter, is still rather shadowy, both like and unlike a musical string. But it is difficult to see where else a writer eschewing mathematics could turn for a way of approaching these strange objects. Nor is there an alternative to the continual shift between macro-analogies, where everyday intuitions can help us understand what is going on, and description of the microworld which string theory actually inhabits. This is a characteristic technique of popular writing about fundamental physics, indeed of much popular science, though the shift is more radical here than with perhaps any previous theory as the scales and forces under discussion are so extreme. It is clear by the end that superstrings are not much like a violin string or a rubber band, but they are pretty much all we have to go on. Now, whatever idea we have built up of superstrings, Greene must build on it to impart some other important properties of these new fundamental entities. In Ogborn's terms, perhaps, he needs to move from creating entities to transforming knowledge.

In fact, one could regard the work of the whole of the rest of the book as a series of transformations of our initial ideas about super-strings, over another two hundred pages of an exposition which, unusually for a work of popular science, get denser and more difficult as it goes on. It would be over-ambitious to comment on even a fraction of the explanatory work these pages undertake. It is worth noting just a few things, though. First, there is real difficulty here in putting meaning into matter, as what matter actually consists of is one of the things at stake in the theory. There are, though, numerous thought experiments. This is partly because, as Greene frequently concedes, there are as yet no actual experiments which might disclose the signature of a superstring. But it is also because there are new thought pictures to be drawn of mathematical ideas. To begin with, there is the small matter of conveying the notion that strings exist in 11 dimensions, instead of 4, to readers whose intuitions are tuned to the 3 spatial and 1 time dimension we can experience at first hand. So there is much contemplation of ants dwelling on the surface of hosepipes, which have one extended linear dimension and one 'curled up' much smaller.

But this homely image gives way quite quickly to attempts to represent six-dimensional shapes, although hosepipes and doughnuts of various configurations constantly reappear as Greene depicts particular topological properties of superstrings and their interactions. Here, the alternation is not so much between the macro and the micro as between the experiential world and a mathematical world which string theory holds is a better representation of the underlying material reality.

We might also discuss the crucial feature of supersymmetry, which is something like normal geometrical symmetry, but neverthe-less, not readily susceptible to a non-mathematical treatment, any more than the concept of particle 'spin' which is closely bound up with it. And there are a host of further conceptual leaps the reader has to contemplate. Strings eventually turn into multi-dimensional membranes, to which the rules that were first introduced when they looked like violin strings still apply, and their properties lead to new ideas about the nature of space and time at the micro-level. By the end of the exposition, the entities sketched in the first chapter on strings have been well and truly transformed, along with most of the readers' notions about the physical world.

All of this suggests, perhaps, that the newest science, this fundamental physics in the making, stretches the proposed vocabulary for describing explanations which I have been discussing to its limits. But this is only to be expected. For it is also stretching explanation to its limits. It may be that Greene's volume takes its place as a landmark attempt to develop a way of depicting superstrings which makes some sense to a non-mathematical reader, with some of his images and ideas adopted by others, some modified, and some supplanted entirely by word pictures drawn by later authors. That will depend on the whole body of theory he discusses becoming established in physics more firmly than it is now, as firmly, say, as quantum mechanics was in the 1920s.

It may also depend on physicists deepening their understanding of the implications of thinking of the most basic entities in the universe as multi-dimensional string-like objects. Greene himself quotes Rutherford's dictum that if you cannot explain some result in physics in simple, non-technical terms, then you do not fully understand its origin, meaning, or implications. On the other hand, John Barrow has argued that it may be precisely when a scientific theory gives rise to mathematically based notions which cannot easily be translated into everyday images – and which, perhaps, do not lend themselves to the kinds of explanatory stories which feature in the majority of popular sciences texts – that we are gaining truly novel understandings of nature. If so, though, one has to ask, will we ever settle for the corollary that there is really no way of explaining them to everyone else?

References

Eger, M. (1993). 'Hermeneutics and the new epic of science'. In *The literature of science: perspectives on Popular Scientific Writing* (ed. M. McRae), pp. 186–212. University of Georgia Press, Athens, GA.

Fleck, L. (1935/1979). *Genesis and development of a scientific fact.* University of Chicago Press.

Greene, B. (2000). *The elegant universe: superstrings, hidden dimensions, and the quest for the ultimate theory.* Vintage, London.

MacIntyre, A. (1977). 'Epistemological crises, dramatic narrative and the philosophy of science'. *The Monist*, 60, 453–72.

Ogborn, J., Kress, G., Martins, I., and McGillicuddy, K. (1996). *Explaining science in the classroom.* Open University Press, Milton Keynes.

Sobel, D. (1996). *Longitude.* Fourth Estate, London.

Stannard, R. (1999). 'Einstein for young people'. In *Communicating science: contexts and channels* (ed. E. Scanlon, E. Whitelegg, and S. Yates), pp. 134–45. Routledge, London.

Turner, M. (1996). *The literary mind: the origins of thought and language.* Oxford University Press.

Turney, J. (1999). 'The world and the world: engaging with science in print'. In *Communicating science: contexts and channels* (ed. E. Scanlon, E. Whitelegg, and S. Yates), pp. 120–33. Routledge, London.

Turney, J. (2001). 'More than story-telling: reflecting on popular science'. In *Science communication in theory and practice* (ed. C. Bryant and S. Stocklmayer), pp. 47–62. Kluwer, Dordrecht.

Weinberg, S. (1993). *Dreams of a final theory.* Hutchinson, London.

Index